EXPERIMENTAL DESIGN FOR AGRO-INDUSTRIES VOLUME II

Jose Luis Ibave González

Rocío Isabel Díaz de la Garza

Joel Badillo Lucero

Edgar Isaac Yáñez Ortíz

2 0 1 9

Experimental Design for Agro-Industries, Volume II

Second Edition

Jose Luis Ibave González, Ph.D.

Rocío Isabel Díaz de la Garza, Ph.D.

M.A. Joel Badillo Lucero

M.A. Edgar Isaac Yáñez Ortíz

José Luis Ibave, Editor

ISBN-13: 978-1-948150-21-7

bsph
Borderland Studies Publishing House

SAMPLING

In most crop experiments in the field or in the greenhouse, yield is determined by harvesting the entire plot or pot. The same is true of many other kinds of measurements. With animals, weight gain may be taken for each animal in the experiment. In food science we may measure the protein content or crude fiber of all of the plant material harvested from a plot. However in some situations or for some measurements it may be difficult or impossible to measure an entire plot.

In forage experiments we may not find it possible to take measurements on the entire plot, but instead will take a certain number of random, square meter, samples in each plot and obtain our measurements from these samples. In weed experiments, commonly a weed count is not taken on the whole plot, but instead, a certain number of small sample plots are selected at random in each main plot and weed counts made on these smaller units. In potato experiments, yield measurements are obtained from the entire plot, but in determining dry matter content, we generally take a random sample of ten pounds of tubers from each plot and determine the dry matter content of this ten pound sample. The same method is commonly used to determinate sucrose content of sugar beets. You can probably think of many other situations where measurements would be made on samples rather than on the entire plot.

Sampling is a procedure of selecting sampling units in such a way that they best represent the population from which they come. There will, of course, be variation from sample to sample within a plot and so the validity of conclusions derived from sample data depends on whether the samples are representative of the plots from which they were taken. The best way to assure that the samples are representative and not giving a biased picture is to have a random selection of samples.

The objective of sampling is to obtain as close an estimate as possible of the measure, which we would have obtained, within the limits of experimental error, had the produce of the entire plot been measured (counted, weighed, etc.).

Choice of Sampling Units

The choice of sampling units will depend upon:

(a) The ease with which they may be obtained and
(b) The nature of the crop or the experimental material

In crops of easily identified individual units such as potatoes, tomatoes, sugar beets, etc., one or more randomly selected individual plants or roots may serve as the sampling unit. With non-spaced crops such as small grains or forage crops, where individual plants may be difficult to identify, the sampling unit may consist of the produce from a meter or half-meter of row, or it may consist of the produce from a meter square or half-meter square. Some measurements, such as counting of weeds, insects, tillers, etc., generally take more time than harvesting a sampling unit. In these cases the size of the sampling unit or the number of sampling units can be reduced.

To simplify the analysis it is advisable to have the same number of sampling units in each plot.

Sources of Variability

In agricultural experiments or for that matter experiments in any field of study two sources of variability must be considered, the variation between plots and the variation within plots. Where sampling is used for the determination of yield or some measurement the variation between sampling units within plots represents the magnitude of the sampling error.

For example, sampling may be used to estimate the yield of three varieties of a crop, each replicated four times. Sampling might consist of ten one-meter-square sampling units taken from each plot. The key-out would be as follows:

Source of Variation	DF
Replicates	3
Varieties	2
Error (plots within replicates)	6 = a
Total for plots	11
Samples within plots	108 = b
Total samples	119

For comparisons between varieties the mean square for error "a" (plots within replicates) must be used. The mean square for samples within plots or error "b" furnishes an estimate of the sampling variance which can be used primarily to determinate the increase in precision which could be obtained in future experiments.

In replicated experiments we are frequently interested in quality factors such as moisture content, dry matter, sugar content, etc. as well as such factors as yield. The quality factors are usually evaluated in a laboratory. Generally we take samples of the harvested plots for the laboratory analyses. It is important to remember that the produce from each plot must be sampled individually because there is likely to be variation between plots receiving the same treatment. Also, at least two sampling units must be taken from each plot if we are to obtain a measure of the variation from sample to sample. In the laboratory analysis, if we want to have a measure of the variation due to analysis then more than one analysis must be run on each sample. Below he have an example of how these different sources of variation compare whit one another.

In an experiment with sugar beets, the mean squares for the percentage sucrose an the apparent purity were as follows:

Sources of Variation	Mean Squares	
	%sucrosa	purity
Plots within replicates	0.86	12.09
Sampling units within plots	0.34	2.87
Lab. analyses within sampling units	0.05	1.21

In general, variability (mean-square) decreases progressively from plots within replicates, to sampling units within plots, to laboratory analyses within sampling units, as it does in this example.

The mean square for "lab. analyses within sampling units" is due to errors in the laboratory procedure. The mean square due to variation between "sampling units within plots" is that due both to laboratory errors ant the true variation between individual sampling units from the same plots. the mean square for "plots within replicates" is due to the laboratory errors, sampling errors and the true residual variation between the different plots. This last mean square is the experimental error used to test the significance of difference between treatments.

Let us compare the mean square in the above table. Because of the larger mean square for "sampling units within plots" as compared with the mean square for "lab. analysis within sampling units" it would be better to take more sampling units from a plot rather than to make may analyses on a single sampling unit. However, it is obvious that the major source of variation is due to the difference between plots. Therefore the number, size and uniformity of these plots is of the most importance. Any marked reduction in experimental error can be expected trough the reduction of this plot variability. This can be accomplished most effectively by an increase in the number of replicates.

An experiment to study seed yield of sweet clover was conducted in Minnesota. Six varieties were grown in three replicates, making a total of eighteen plots. A measure of the yield of seed was obtained by taking four one-meter-square samples from each plot. Seed yields are shown in Table I.

TABLE I
Sweet Clover Seed Yields

	Variety						
	A	B	C	D	E	F	Total
Rep.1	20	24	14	28	22	27	
	22	29	19	24	25	31	
	24	26	16	28	22	29	
	24	26	18	25	23	29	
	90	105	67	105	92	116	575
Rep.2	18	26	11	28	21	31	
	22	30	14	25	23	34	
	19	26	11	26	21	31	
	22	28	13	27	21	34	
	81	110	49	106	86	130	562
Rep. 3	24	23	18	28	21	25	
	27	26	21	31	20	27	
	24	23	20	28	22	28	
	24	25	18	30	20	28	
	99	97	77	117	83	108	581
Total	270	312	193	328	261	354	1718

The calculation of the sum of squares is as follows:

Correction Factor $= 1718^2/72 = 40{,}993.3889$

$S\ (x)^2 = (20^2 + 22^2 + 24^2 + \ldots + 28^2 + 28^2) =$
$42{,}783.0000 - CF = 1{,}788.6111$

$S\ (x\ rep)^2/24 = (575^2 + 562^2 + 581^2)/24 =$
$41{,}001.0000 - CF = 7.8611$

$S\ (x\ var)^2/12 = (270^2 + 312^2 + \ldots + 354^2)/12 =$
$42{,}376.1667 - CF = 1{,}382.7778$

$S\ (x\ rxv)^2/4 = (90^2 + 105^2 + \ldots + 108^2)/4 =$
$42{,}633.5000 - CF - S_r - S_v = 249.4722$

TABLE II

Analysis of Variance

Source of Variation	DF	SS	MS	Components of Variance
Replicates	2	7.8611	3.9306	$s^2 + 4s^2vr + 24s^2r$
Varieties	5	1,382.7778	276.5556	$s^2 + 4s^2vr + 12s^2v$
Plots within reps. (experimental error)	10	249.4722	24.9472	$s^2 + 4s^2vr$
Plot totals	17	1,640.1111		
Sampling units within plots	54	148.5000	2.7500	s^2
Total for samples	71	1,788.6111		

From the mean squares and the expressions for the components we can solve for the numerical value of the estimated components:

$s^2 = 2.7500$

$s^2vr = (24.9472 - 2.7500)/4 = 5.5493$

$s^2v = (276.5556 - 24.9472)/12 = 20.9674$

$s^2r = (3.9306 - 24.9472)/24 = -0.8757$

Now, with respect to sampling the question is, could we improve the experiment more rapidly by increasing the number of sampling units or by increasing the number of replications? We can answer this question by comparing the estimates of variance components which we calculated above.

In the original analysis the error variance of a mean equals 24.9472/12 = 2.0789. The divisor is 12 because each variety mean comes from three plots with 4 samples from each plot. In terms of variance components this can be calculated as follows:

1/reps. $(s^2vr + s^2/\text{samples}) = 1/3(5.5493 + 2.75/4) = 2.0789$ as before

Suppose we had used four replicates instead of three. Then substituting in the equation we have:

$1/4(5.5493 + 2.75/4) = 1.5592$ error variance of a variety mean.

Or, suppose we had taken ten samples per plot instead of six. Then substituting in the equation we have:

$1/3(5.5493 + 2.75/10) = 1.9414$ error variance of a variety mean.

Using this method of estimation we can set up table III for various combinations of number of replicates and number of sampling units.

TABLE III

Estimated Error Variance of Variety Means

Number of samples	Number of Replicates 3	4	5	6	7	8	s^2vr + s^2/no. samples
2	2.3081	1.7311	1.3849	1.1540	.9892	.8655	6.9243
4	2.0789	1.5592	1.2474	1.0395	.8910	.7796	6.2368
6	2.0025	1.5019	1.2015	1.0013	.8582	.7510	6.0076
8	1.9643	1.4732	1.1786	.9822	.8418	.7366	5.8930
10	1.9414	1.4561	1.1649	.9709	.8320	.7280	5.8243

From the Table III we see that if the relative sizes of s^2 and s^2vr remained the same in future experiments then we could reduce the error variance of the variety means more rapidly by increasing the number of replicates than by increasing the number of sampling units per plot. It is possible to reduce the variance of the mean to any point of precision desired provided enough replicates and/or samples are taken. The above technique can be very useful in planning futures experiments. Notice carefully, that for the technique to be truly effective we must have a very good estimate of s^2 and s^2vr.

To repeat, in general if the sampling mean square were very large we would consider increasing the number of sampling units. If the experimental error mean square were large (as in this case) we would increase the number of replicates and, if need be, reduce the number of sampling units because this would give us the greatest reduction in error.

COMPARISON OF MEANS

"a priori" and " a posteriori"

When an experiment has been completed and the analysis of variance conducted the next step is, almost always, to determinate which treatments differ from which other treatments. To facilitate the discussion on comparison of means we will use data from an actual experiment. This experiment was designed to study the effect of different sugars, added to the tissue culture medium, on the growth of stem sections of pea plants. The complete data is shown in Table I.

TABLE I

Effect of Sugars on Stem Growth of Pea

	Control	2% Glucose	2% Fructose	1% Glucose +1% Fructose	2% Sucrose
	75	57	58	58	62
	67	58	61	59	66
	70	60	56	58	65
	75	59	58	61	63
	65	62	57	57	64
	71	60	56	56	62
	67	60	61	58	65
	67	57	60	57	65
	76	59	57	57	62
	68	61	58	59	67
Total	701	593	582	580	641
Mean	70.1	59.3	58.2	58.0	64.1

Grand Total 3097

The calculation of the sums of squares and the analysis of variance are shown below and in Table II.

Correction Factor = $(3097)^2/50 = 191{,}828.18$

$\text{Sum}(x)^2 = (75^2 + 67^2 + ... + 62^2 + 67^2) - \text{CF} = 1322.82$

$\text{Sum}\ (x)^2\ \text{treat.} = (701^2 + ... + 641^2)/10 - \text{CF} = 1077.32$

TABLE II

Analysis of Variance

Source of Variation	DF	SS	MS	F
Treatment	4	1077.32	269.33	49.33**
Error	45	245.50	5.46	
Total	49	1322.82		

Comparison of Means

Basically there are two ways in which tests can be carried out. In the first case the tests or comparisons which are to be made are chosen or designated independently of the results of the experiments. That is, they are planned before the experiment has been conducted and results obtained. Such tests or comparisons are called planned or "a priori" comparisons. In the above experiment we would undoubtedly have planned to test the control against the four treatments with added sugar to answer the question : Does the addition of sugars have an effect on the growth of stem section? We would also probably want to test for difference among sugars. One such test would be pure sugars (glucose, fructose, sucrose) versus the mixed sugar treatment. Such tests may be conducted regardless of the results of the analysis of variance. If the analysis of variance did not show significant difference between treatments we could still carry out our planned or "a priori" tests.

In the second case we conduct the experiment, carry out the analysis of variance and then examine treatment means. It is at this point that we generally decide what test to conduct, what comparisons to make. Usually we will compare means which are markedly different. For example, sucrose with a mean of 64.1 appears to have had less of a growth inhibiting effect than fructose with a mean of 58.2. We might therefore decide to test whether there is indeed a comparisons, which suggest themselves as a result of the completed analysis, are called unplanned or "a posteriori" comparisons. Now, these unplanned tests should be carried out only if the analysis of variance shows significant differences between treatments. The possible number of such comparisons include comparisons between all possible pairs of means. When there are "t" treatment means, there can be t(t-1)/2 possible comparisons between means. In our experiment there are 5(4)/2 = 10 possible comparisons.

The reason we make a distinction between "a priori" and "a posteriori" comparisons is that the tests of significance appropriate for the two types of comparisons are different.

"a priori" Comparisons

We will first discuss and compute " a priori" comparisons, those planned before the experiment is carried out. The general rules for making planned comparisons are really very simple and we will use the above stem section experiment to demonstrate these.

Our first comparison would undoubtedly be between the control treatments and the four treatments in which sugar was added. There will thus be two groups, the control groups with a total of 701 based on ten observations and the "sugars" group with a total of 2396 based on forty observations. The sum of squares is calculated as follows:

$$SS(\text{control vs. sugars}) = [(701)^2/10 + (593+582+580+641)^2/40] -$$

$$(701+593+582+580+641)^2/50 = 832.32$$

In this case the correction term $(3097)^2/50$ is the same as for the analysis of variance because all five treatments are involved. A comparison between two groups has only 1 degree freedom and so the sum of squares = mean square. This mean square is tested against the error mean square of the analysis of variance.

$$F = \frac{MS(\text{control vs. sugar})}{MS\ \text{error}} = \frac{832.32}{5.46} = 152.44^{**}$$

The test is highly significant (0.01 level) showing that the addition of sugars significantly retarded the growth of the stem sections.

Next we will test whether the effect of the mixture of sugars (1% glucose + 1% fructose) differs significantly from the effect of the three pure sugars.

$$SS(\text{mixed sugars vs. pure sugars}) = [(580)^2/10 + (593+582+641)^2/30]$$

$$- (593+582+580+641)^2/40 = 48.13$$

In this case the correction factor is different from that used in the analysis of variance since it is based on the sum of the sugar treatments only, the control treatment is not include. The appropriate test is:

$$F = \frac{MS(\text{mixed sugars vs. pure sugars})}{MS\ \text{error}} = \frac{48.13}{5.46} = 8.82^{**}$$

The test is highly significant showing that the reduction in growth of stem sections was significantly greater with a mixture of glucose and fructose than it was with pure sugars.

A final planned test would probably be the comparing the three pure sugars. Notice that this comparison has two degree of freedom since it is based on three means.

$$SS(\text{among pure sugars}) = [(593)^2/10 + (582)^2/10 + (641)^2/10]$$
$$- (1816)^2/30 = 196.87$$

$$MS(\text{among pure sugars}) = 196.87/2 = 98.435$$

$$F = \frac{MS(\text{among pure sugars})}{MS\ \text{error}} = \frac{98.435}{5.46} = 18.06^{**}$$

The test is highly significant showing that the three pure sugars are significantly different among themselves. This is probably due to the sucrose which has a markedly higher mean. Let us assume that the above three test were the ones and the only ones which were planned in advance. Notice carefully then we cannot test the sucrose treatment against the other two pure sugar treatments (glucose and fructose), because that would be an "a posteriori" test which suggested itself to us after we had examined the results.

Our "a priori" tests could have been quite different depending on the original objectives of the experiment. In setting up the experiment we might have been interested in testing (1) control vs. sugars, (2) disaccharides (sucrose) vs. monosaccharides (glucose, fructose, glucose + fructose), (3) mixed vs. pure monosaccharides and finally (4) glucose vs. fructose. Each of these mean squares has 1 degree freedom, totaling 4 degree of freedom, the same as treatment degrees of freedom.

We must be careful in our approach to "a priori" testing. It would not be valid to decide "a priori" that one wanted to test every treatment against every treatment, that is $t(t-1)/2 = 10$ comparisons. For t groups the sum of the degrees of freedom of the separate "a priori" test should not exceed t-1, in our case 4.

In our sugar experiment, the three planned test we conducted are independent of one another and therefore the three sum of squares based on 1, 1 and 2 degrees of freedom add up to the sum of squares among treatments of the original analysis of variance based on 4 degrees of freedom.

	SS	DF
control vs. sugars	832.32	1
mixed vs. pure sugars	48.13	1
among pure sugars	196.87	2
among treatments	1,077.32	4

Not all sets of comparisons yield sums of squares that add up to the treatment sum of squares. For example, another comparison that could legitimately be made in the above experiment is glucose and fructose considered together against the mixture of glucose and fructose.

SS(glucose and fructose vs. glucose-fructose mixture) =

$$[(593+582)^2/20 + (580)^2/10] - (593+582+580)^2/30 = 3.75$$

$$F = \frac{\text{MS(of above)}}{\text{MS error}} = \frac{3.75}{5.46} = 0.69$$

The test is clearly not significant indicating that glucose and fructose have the same effect on growth whether applied alone or in a mixture.

This 1 DF comparison although leading to a legitimate test cuts across the comparisons made previously and therefore it is not independent of those comparisons. The comparison of mixed vs. pure sugars tested almost the same thing. Also, if we include this comparison with the others then we have partitioned the degrees of freedom among treatments into 1+1+2+1 = 5 degreesof freedom. However there are only 4 degrees of freedom in the treatment component. When the comparisons can be so added that the sum of the degree of freedom and the sum of the sum of squares the partitioning is called orthogonal. When the comparisons cannot be so added the partitioning is called non-orthogonal. Both can be used but it is important to realize and understand exactly what you are doing.

"a posteriori" Comparisons

The development of proper methods for testing "a posteriori" hypotheses in the analysis of variance has been worked on by a number of statisticians. Names and tests you have become familiar with o will become familiar with are: L.S.D., Tukey's "w" test, Duncan's Multiple Range Test, Student-Newman-Keuls test, Scheffer's test, etc. Some of the tests like L.S.D. and Tukey's test use a single value. Some like Duncan's test use increasing values depending on the range. In the next handout we will cover a few of these method.

Test to Determinate Which Treatment Means Differ Significantly From Which Other Means

There are a number of tests which are used or have been used in the past to determinate which treatments are significantly different.

1. The "t" test is of course used when are comparing two means. Most commonly it is used in experiment in which there are only two treatments and if the "t" test shows significance this automatically indicates that the two treatment means differ significantly.

2. When an experiment contains more than two treatments and the F test indicates significant differences between treatments, this does not tell us which means differ from which others. We, therefore, have to go one step further to answer this question. Let us assume a randomized complete block experiment of nine treatments with an analysis of variance as shown in Table I.

TABLE I

Analysis of Variance

Source of Variation	DF	SS	MS	F
Replicates	3	9,158.5715	3,052.8572	7.90**
Treatments	8	44,169.9048	5,521.2381	14.29**
Error	24	9,273.9048	386.4127	
Total	35	62,602.3811		

Treatment means were as follows, arranged in order from lowest to highest yield:

Treatment	Mean Yield
G	150.5
C	175.8
F	181.3
D	195.5
A	199.4
E	218.2
H	223.6
B	228.8
I	240.5

Now we will use this data to demonstrate four different methods of determining the significance of the difference between means. Throughout we will use the 5% level of significance even though the above test for treatments was significant at the 1% level.

(a) L.S.D. = LEAST - SIGNIFICANT DIFFERENCE

The LSD was widely used in the past but has been superseded in many, many cases by newer more valid tests which are discussed later.

Now a point to remember: if the F value for treatments is not significance the evidence is against rejecting the null hypothesis and we say treatments do not differ significantly. In most cases we must stop here and must not make treatments comparisons. Exceptions do occur when certain comparisons were planned in the beginning. A valid test for planned comparisons of paired means is the LSD. The 5% level is given by the following equation:

$$LSD_{(.05)} = t_{(.05)}\ sd^{-}$$

$t_{(.05)}$ is the tabular value of t for the error degrees of freedom.
sd^{-} is the standard error of the mean difference $= \sqrt{2s^2/r}$
where s^2 is the error variance and r the number of observation per mean.

For our experiment

$t_{.05}$ for 24DF = 2.064

$sd^{-} = \sqrt{2 \text{ X } 386.4127 / 4} = 13.8999$

$LSD_{(.05)} = 2.064 \text{ X } 13.8889 = 28.7$

Let us compare treatment means using LSD. The treatment means are ranked below.

Treatment	Mean Yield	
G	150.5	a
C	175.8	ab
F	181.3	b
D	195.5	bc
A	199.4	bc
E	218.2	cd
H	223.6	cd
B	228.8	d
I	240.5	d

Now to determinate the difference we take G + LSD = 150.5 + 28.7 = 179.2. Treatment C is less than 179.2 but treatment F is greater. Therefore, C does not differ from G but F and all others do. Next we take C + LSD = 175.8 + 28.7 = 204.5. Treatments F, D and A are less than 204.5 but E is greater. Therefore C does not differ from F, D and A but does differ significantly from E and the rest.
Treatment means followed by a same letter do not differ from one another.

(b) DUNCAN'S NEW MULTIPLE RANGE TEST

The tends to be too "weak" a test in that if we are comparing a large number of treatments, say nine,the LSD gives us too many significant differences. When comparing each treatment mean with every another treatment mean one solution to the problem is to develop a method calling for a set of significant differences of increasing size, the size to depend upon the closeness of the means and the largest for the extremes. Duncan has developed the new multiple range test to compare each treatment mean with every other treatment mean.

The 5% multiple range test is illustrated here for the above data. The procedure is:

1. Determinate $s\bar{x} = \sqrt{\text{error mean square} / r}$

$$= \sqrt{386.4127 / 4} = 9.8287$$

2. In most statistic texts you will find a table "Significant Studentized Ranges for 5% and 1%. New Multiple-Range Test". For the 5% multiple range test enter the table at row n_2= 24 (error degrees of freedom) and extract the significant studentized ranges (SSR) for p=2 to 9, the sizes referring to the number of means involved in a comparison. The SSR are then multiplied by $s\bar{x}$ to give the least significant ranges (LSR) as shown in the table below.

$s\bar{x}$ = 9.8287

Value of p	SSR	LSR
2	2.92	28.7
3	3.07	30.2
4	3.15	31.0
5	3.22	31.6
6	3.28	32.2
7	3.31	32.5
8	3.34	32.8
9	3.37	33.1

Let us now compare treatment means using Duncan's new multiple range test. The treatment means are ranked below:

Treatment	Mean Yield	
G	150.5	a
C	175.8	ab
F	181.3	b
D	195.5	bc
A	199.4	bcd
E	218.2	cde
H	223.6	cde
B	228.8	de
I	240.5	e

This table is exactly the same as the one obtained using LSD except for one difference. With LSD, A and B were significantly different; with Duncan's test they are not. Often times the differences between the two results will be more marked. Notice the LSR for p=2 is the same as the LSD = 28.7.

Now how do we go about comparing treatment means.

Start with treatment G and compare it with C. We use p=2. 150.5 + 28.7 = 179.2, so G and C do not differ.

To compare G and F we use p = 3. 150.5 + 30.2 = 180.7, so G and F do differ.

Now go to treatment C and compare it with F. We use p=2. 175.8 + 28.7 = 204.5; c and F do not differ.

To compare C and D we use p = 3. 175.8 + 30.2 = 206.0 C and D do not differ.

To compare C and A we use p = 4. 175.8 + 31.0 = 206.8 C and A do not differ.

To compare C and E we use p = 5. 175.8 + 31.6 = 207.4 C and E do differ.

We do this same procedure for treatment F, D, A, etc.

Duncan's new multiple-range test has been widely used and is still used by many research workers. However, it has been questioned rather recently by a number of statisticians including Duncan himself. Duncan's test like the LSD tends to show significant differences too often.

(c) TUKEY'S "W" PROCEDURE

Also called the honestly significant difference (HSD) procedure. This procedure is similar to the LSD in that it requires a single value for judging the significance of all differences.

The w procedure consists of computing

$$w = q^{\alpha}\ (p, n_2) s\bar{x}$$

Where $\mathbf{q}^{\alpha}$ for either = .05 or .01 is obtained from the table "Upper Percentage Points of the Studentized Range", which can be found in the appendix.

p is the number of treatments, $\mathbf{n_2}$ is the error degrees of freedom

w is used to judge the significance of each of the observed differences.

For our experiment

$w = q^{\alpha}\ (p, n_2)\ s\bar{x}$, $p = 9$, $n_2 = 24$

$q_{.05}$ for $p = 9$ and $n_2 = 24$ is 4.81

$s\bar{x} = 9.8287$

$w_{.05} = 4.81 \times 9.8287 = 47.3$

Let us now compare treatment means using Tukey's w procedure:

Treatment	Mean Yield	
G	150.5	a
C	175.8	ab
F	181.3	abc
D	195.5	abcd
A	199.4	bcd
E	218.2	bcd
H	223.6	cd
B	228.8	d
I	240.5	d

If we compare this table with Duncan's new multiple-range test we see that we have indeed found fewer significant differences using Tukey's w procedure.

(d) STUDENT-NEWMAN-KEUL'S TEST

This test is also called the Newman-Keul's test or simply Keul's method. Like Duncan's procedure it is a multiple-range test.

The procedure is to compute

$$Wp = q^{\alpha}(p, n_2)\ s_{\bar{x}}$$

where q^{α} is obtained as in Tukey's w procedure but it is obtained for not just one p but for the range of p's covered in the experiment.

n_2 is the error degrees of freedom.

For our data we obtain the fallowing table:

$s_{\bar{x}} = 9.8287$

p	q^{α} (p, n_2)	Wp
2	2.92	28.7
3	3.53	34.7
4	3.90	38.3
5	4.17	41.0
6	4.37	43.0
7	4.54	44.6
8	4.68	46.0
9	4.81	47.3

Notice that our last value for p = 9 is 47.3. This is the single value required for Tukey's w procedure.

We will now compare the treatment means using Newman-Keul's test.

Treatment	Mean Yield	
G	150.5	a
C	175.8	ab
F	181.3	abc
D	195.5	bcd
A	199.4	bcd
E	218.2	cde
H	223.6	de
B	228.8	de
I	240.5	e

If we compare this test with the others we find that the Newman-Keul's test gives slightly fewer significant difference than Duncan's but more than Tukey's procedure.

COMPARING ALL MEANS WITH A CONTROL

Sometimes in an experiment we are not interested in comparing treatment means but is determining which means differ from a control. For such a set of comparisons, that is, control against each of the other treatment means, the LSD is not wholly valid. Dunnett has developed a valid procedure. This procedure requires a single difference for judging the significance of observed difference, and tables are available for one- and two- sided comparisons.

Let us apply this procedure to our data assuming that the treatment D is a control. The procedure is as follows:

Consider the means as a set of p treatments means and a control such that p comparisons are to be made. In this example the comparisons is to be two-tailed with error rate .05. We enter the table "Table of t for Two-Sided Comparisons Between p Treatment Means and a Control for a Joint Confidence Coefficient of p=.95". We obtain a "t" value from the table for p=8 and error degrees of freedom = 24 (This is not Student's t). Our value is 2.96

The significant difference is given by the following equation:

$$d' = t(\text{Dunnett})\ s_{\bar{d}}$$

$$= 2.96 \times 13.8999 = 41.1$$

Let us compare each of our treatment means with the control (treatment D).

Treatment	Mean Yield	
G	150.5	b
C	175.8	a
F	181.3	a
D (control)	195.5	a
A	199.4	a
E	218.2	a
H	223.6	a
B	228.8	a
I	240.5	b

We find that treatments G and I differ significantly from the control. Compare this with the other tests and see what you find.

MISSING PLOT TECHNIQUE

Accidents of one kind or another often cause the loss of experimental data. An animal may die, one or more plots may be seriously damaged, errors may be made when recording data or data may be lost. A missing plot nullifies the addition theorem of the sums of squares. However, we can estimate a value for the missing plot and enter this value in the data table. We then proceed to carry out the analysis of variance in the usual way. This will give us an approximate analysis of variance which can be made exact by correction of a bias in the sum of squares for treatments.

We will go trough the steps involved in estimating the values for missing data, first when there is only one value missing and then when there are two values missing. e will show the steps involved with randomized complete block design. Missing values can also be calculated for other designs such as latin square and the split-plot. We won't show calculations for these designs but most statistic test will show the procedures.

The data from a potato trial is shown in Table I. Let us assume that the yield record for treatment E, replicate 2 (210 cwt.) is missing. The treatment total then becomes 915-210 = 705, the replicate total becomes 1756 - 210 =1546 and the grand total becomes 7260 - 210 = 7050.

TABLE I
Potato Fertilizer Trial (yield in cwt./acre)

	Replicates				
Treatment	1	2	3	4	Total
A	164	146	136	147	593
B	191	142	130	157	620
C	201	152	138	151	642
D	191	180	178	183	732
E	278	210(228)	205	222	915(705)(933)
F	248	239	209	226	922
G	230	210	160	203	803
H	267	243	230	231	971
I	298	234	267	263	1062
Total	2068	1756 (1546) (1774)	1653	1783	7260 (7050) (7278)

The equation for calculating a missing plot is:

$$X = \frac{aT + bR - S}{(a-1)(b-1)}$$

where a = number of treatments = 9
b = number of replicates = 4
T = Sum of the treatments with the missing datum = 705
R = Sum of the replicates with the missing datum = 1546
S = Sum of all the recorded data = 7050

$$X = \frac{9(705) + 4(1546) - (7050)}{(9-1)(4-1)} = \frac{6345 + 6148 - 7050}{24}$$

$$= 228.3 = 228$$

The calculated figure, 228, is placed in the table for treatment E, replicate 2. The treatment total is now 933, the replicate total is now 1774 and the grand total is now 7278. The calculation of sum of squares is as follows:

Correction Factor $= (7278)^2/36 = 1{,}471{,}369.0000$

$S(x)^2 = 1{,}544{,}300 - CF = 72{,}931.0000$

$S(x\ rep.)^2/n = 1{,}481{,}688.6667 - CF = 10{,}319.6667$

$S(x\ Treat.)^2/n = 1{,}530{,}026.0000 - CF = 58{,}657.0000$

The analysis of variance proceeds as usual with two modifications.

(1) The degree of freedom for total and for error are each decreased by 1.
(2) The sum of squares for treatments is decreased to correct for bias.

Correction for treatment bias $= \frac{[R - (a-1)X]^2}{a(a-1)}$

$$= \frac{[1546 - (9-1)228]^2}{9(9-1)} = \frac{(278)^2}{72} = 1{,}073.3889$$

Corrected sum of squares for treatments =

58,657.0000 - 1,073.3889 = 57,583.6111

TABLE II
Analysis of Variance

Source of Variation		DF	SS	MS	F
Replicate		3	10,319.6666	3,439.8888	20.01**
Treatment		8	57,583.6111	7,197.9513	41.87**
Error	24-1	23	3,954.3334	171.9275	
Total	35-1	34	72,931.0000		

As we see there are significant differences between treatments. One point of interest, compare the true plot yield which was actually obtained, 210 cwt., with the calculated value assuming a missing plot, 228 cwt. You will notice that there is a fair difference in the two values. So the calculated value is not a measure of what would have been obtained but is a measure of what should have been obtained.

Now let us assume that the yield records for two plot are missing, treatment E in replicate 2 and treatment C in replicate 4. The data are shown in Table III. In this case, we have left the missing plot blank and have put in the treatment totals, replicates total and grand total as they would appear if the data were actually missing.

TABLE III
Potato Fertilizer Trial With Two Missing Plots

	Replicate				
Treatment	1	2	3	4	Total
A	164	146	136	147	593
B	191	142	130	157	620
C	201	152	138		491
D	191	180	178	183	732
E	278		205	222	705
F	248	239	209	226	922
G	230	210	160	203	803
H	267	243	230	231	971
I	298	234	267	263	1062
Total	2068	1546	1653	1632	6899

Step 1. We will fist estimate the missing value for treatments C, replicate 4.But we must first run a dummy value for treatment E replicate 2. Let us assign a value of 200. This make our grand total 6899 + 200 = 7099

Step 2. Calculate a value for the missing treatment C, replicate 4.

$$x = \frac{aT + bR - S}{(a-1)(a+1)} \qquad a = 9 \quad b = 4 \quad T = 491 \quad R = 1632$$

S = sum of all observed values
(including the dummy value for treatment E, replicates 2) =7099

$$x = \frac{9(491) + 4(1632) - 7099}{(9-1)(4-1)} = 160.33 = 160$$

Step 3. Assign the value of 160 to treatment C, replicate 4. Also add 160 to the grand total 7099 + 160 = 7259.

Step 4. Remove the dummy value we ran in for treatment E, replicate 2 of 200 from the total 7259 - 200 = 7059.

Step 5. Calculate a value for the missing treatment E, replicate 2.

$$x = \frac{9(705) + 4(1546) - 7059}{(9-1)(4-1)} = 227.9 = 228$$

Step 6. Assign new value of 228 to treatment E, replicate 2. Also add 228 to the grand total 7059 + 228 = 7287.

Step 7. Recalculate the value for treatment C, replicate 4. First we subtract the first estimated value of 160 from the grand total = 7287 - 160 = 7127.

$$x = \frac{9(491) + 4(1632) - 7127}{(9-1)(4-1)} = 159.2 = 159$$

Our original estimated value for treatment C, replicate 4 was 160 cwt. The revised estimate is 159 which is very close to 160. We would not expect any change if we were to re-estimate treatment E, replicate 2 and so we stop here. If the second estimate of treatment C, replicate 4 had differed from our estimate for treatment E, replicate 2. Our estimated values for the two missing plot values are:

Treatment C, replicate 4 = 159

Treatment E, replicate 2 = 228

The new table of yield with the estimated values is shown in Table IV.

TABLE IV
Potato Fertilizer Trial With Two Estimated Values

	Replicate				
Treatment	1	2	3	4	Total
A	164	146	136	147	593
B	191	142	130	157	620
C	201	152	138	159	650
D	191	180	178	183	732
E	278	228	205	222	933
F	248	239	209	226	922
G	230	210	160	203	803
H	267	243	230	231	971
I	298	234	267	263	1062
Total	2068	1774 1546	1653	1632	7286 6899

As before, when we had only one missing plot value we must make a correction for bias as follows:

Step 8. We first obtain the sum of squares for Table IV:

Correction Factor = $(7286)^2/36$ = 1,474,605.4444

$S(x)^2$ = 1,546,780.0000 - CF = 72,174.5556

S (x rep.)/n = 1,484,865.5556 - CF = 10,260.1112

S (x treat.)/n = 1,532,610.0000 - CF = 58,004.5556

Error SS = 72,174.5556 - 10,260.1112 - 58,004.5556 =
= 3,909.8888

The only figure from the above calculations which we shall use is the error sum of squares = 3,909.8888 with 24-2 = 22 degrees of freedom.

Step 9. Next we calculate the total sum of squares for Table IV, excluding the missing values

$$\text{Correction Factor} = (6899)^2/34 = 1,399,888.2647$$

$$s(x)^2 = 1,469,515.0000 - CF = 69,626.7353$$

Then we calculate the replicate sum of squares, excluding the missing values.

$$S(xrep.)^2/n = (2068)^2/9 + (1546)^2/8 + (1653)^2/9 + (1632)^2/8 =$$

$$1,410,473.9444 - CF = 10,585.6797$$

Step 10. The unbiased estimate of the treatment sum of squares is obtained as follows:

		Sum of squares
Total, excluding missing values		69,626.7353
Replicate,excluding missing values	10,585.6797	
Error, complete data	3,909.8888	
	14,494.5685	
Treatment sum of squares, by subtraction		14,494.5685
		55,131.1668

Step 11. The final analysis of variance is shown in Table V

TABLE V
Analysis of Variance

Sources of Variation	DF	SS	MS	F
Treatment	8	55,131.1668	6,891.3958	38.78**
Error	22	3,909.8888	177.7222	

At the beginning of the exercise, where we had only one missing value, we calculated the correction for bias by means of a simple formula. We could have also corrected for bias by following steps 8, 9 and 10, which we shall now do.

Error SS = 3,954.3334

Correction Factor, excluding missing plot= $(7050)^2/35$= 1,420,071.4285

$s(x)^2$ = 1,492,316.0000 - CF = 72,244.5715

$S(\text{xreps.})^2/n$ excluding missing value = $(2068)^2/9 + (1546)^2/8 + (1653)^2/9 + (1783)^2/9$ = 1,430,778.0555 - CF = 10,706.6270.

The unbiased estimate of the treatment sum of squares equals:

		Sum of squares
Total, excluding missing values		72,244.5715
Replicate, excluding missing values	10,706.6270	
Error, complete data	3,954.3334	
	14,660.9604	
Treatment sum of squares		14,660.9604
by subtraction		57,583.6111

If you go back to the middle of page 2 you will see that this value 57,583.611, is the same as the one we obtained by the short method.

CHI-SQUARE (χ^2) TEST

Up to now we have been using the mean and the standard deviation to demostrate and test differences and make comparisons: an observed against a hypothetical distribution, one observed distribution against another, a comparison of treatment means. Generally the data we have been studying has been measured on a continuos scale. However, in many experiments the data are merely classified into groups or categories; success or failure, infected or not infected, etc. Usually the data consists of the number of individuals falling into each of the specified categories. In experiments of this type the statistical analysis is carried out by means of the Chi-square test, the formula for which is:

$$\chi^2 = \text{Sum}\ \frac{(\text{observed} - \text{calculated})^2}{\text{calculated}} = \text{Sum}\ \frac{(O - C)^2}{C}$$

where O = the number observed in any one category or group.
C = the theoretically calculated number for the same category or group. The distribution of the calculated numbers is adjusted to add up to the same total as the total number of observations.

The Chi-square test is most widely used to:

a) Test an observed distribution against one expected on the basis of some hypothesis. Called a test for "goodness of fit".

b) Individuals may be classified for two characteristics (resistance to two diseases) and we want to test for association or independence of the two characters. Called a test for "independence".

GOODNESS OF FIT

This test is used a great deal in genetics where we commonly want to find the number of factors involved in the inheritance of a character or determinate the segregation of a specific cross. The goodness of fit test is, of course, also used in many other situations where we are comparing observed and expected distribution.

<u>Problem</u> <u>1</u> The progeny of a wheat cross segregrate into 82 tall and 39 dwarf plants in the F2. We want to determinate if the data fits a 3:1 ratio.

		O	C	O-C	$(O-C)^2$	$(O-C)^2/C$
A	Tall	82	90.75	-8.75	76.5625	0.8437
a	Dwarf	39	30.25	8.75	76.5625	2.5310
Total		121	121.00			

χ^2=3.3747

$\chi^2 = 3.3747$ DF = r-1 = 1 χ^2(tabulated) at 5% level = 3.84

Our Chi-square value is not significant at the 5% level, although it is close.

The calculated values were obtained as follows:

121/4 = 30.25 the calculated number of dwarf plants
30.25 x 3 = 90.75 the calculated number of tall plants.

Chi-square for this example can be calculated by a shorter method:

$$\chi^2 = \frac{(A - 3a)^2}{3n} = \frac{(82 - 117)^2}{363} = 3.3746 \text{ as before}$$

For large number of observations, the above formula can be used as is, but if the number is small, or if the number of observations in any category is small a correction for continuity must be used. The correction is always used in such a way as to reduce the size of the Chi-square value. With the first method of calculation with the above example 0.5 is added or substracted from each observed value in such a way to reduce χ^2.

		Uncorrected		Corrected			
	O	C	O-C	O	O-C	$(O-C)^2$	$(O-C)^2/C$
A Tall	82	90.75	-8.75	82.5	-8.25	68.0625	0.7500
a Dwarf	39	30.25	8.75	38.5	8.25	68.0625	2.2500
Total	121	121.00		121.0			

χ^2 = 3.0000

$\chi^2 = 3.0000$ With the uncorrected numbers it was 3.3747

Using the shorter method of calculating χ^2 and applying the correction for continuity:

$$\chi^2 = \frac{[(A - 3a) \pm 2]^2}{3n}$$

The 2 is either added or substracted so as to reduce the size of the Chi-square value.

$$\chi^2 = \frac{[(82 - 117) + 2]^2}{363} = 3.0000 \text{ as before}$$

Generally the correction is used only when the number in any one category is less than ten. Also, the correction is generally used only when the degrees of freedom is 1. The correction can be applied to cases there is more than 1 degree of freedom but the correction is complicated and in such cases the effects of small numbers are usually not serious.

The Chi-square test is applied in the same manner to other simple genetic ratios such as 1:1, 15:1, 9:7. The short method for calculating Chi-square for these ratios is given below:

Ratio A:a	Formula	Corrected Formula
1:1	$\chi^2 = \frac{(A-a)^2}{n}$	$\chi^2 = \frac{[(A-a) + 1]^2}{n}$
15:1	$\chi^2 = \frac{(A-15a)^2}{15n}$	$\chi^2 = \frac{[(A-15a) + 8]^2}{15n}$
9:7	$\chi^2 = \frac{(7A-9a)^2}{63n}$	$\chi^2 = \frac{[(7A-9a) + 8]^2}{63n}$

Problem 2. The F2 progeny of a bean cross, tall yellow (AAbb) x dwarf green (aaBB) segregated into 1949 tall green, 706 tall yellow, 702 dwarf green and 280 dwarf yellow. Using Chi-square test we will determinate if this segregation fits a 9:3:3:1 ratio

	O	C	O-C	$(O-C)^2$	$(O-C)^2/C$
AB Tall, Green	1949	2045.79	-96.79	9368.3041	4.5793
Ab Tall, Yellow	706	681.93	24.07	579.3649	0.8496
aB Dwarf, Green	702	681.93	20.07	402.8049	0.5907
ab Dwarf, Yellow	280	227.31	52.69	2776.2361	12.2134
Total	3637	3636.96			

$X^2 = 18.2330$

$\chi^2 = 18.2330$ DF = 3 χ^2 (calculated at 1% level) = 11.30

The Chi-square is significant at the 1% level. Therefore the data do not fit the 9:3:3:1 ratio. From this the breeder would infer that there is linkage between the genes for plant height and plant color. In this example the calculated values are obtained as follows:

3637/16 = 227.31 the calculated number of dwarf, yellow plants.

227.31 x 3 = 681.93 the calculated numbers of dwarf, green and of tall, yellow plants.

227.31 x 9 =2045.79 the calculated number of tall, green plants

A shorter method of calculation is available for this example:

$$X^2 = \frac{16}{n}\left[\left(\frac{AB^2}{9} + \frac{Ab^2}{3} + \frac{aB^2}{3} + \frac{ab^2}{1}\right)\right] - n$$

$$= \frac{16}{3637}\left[\left(\frac{1949^2}{9} + \frac{706^2}{3} + \frac{702^2}{3} + \frac{280^2}{1}\right)\right] - 3637 = 18.2328 \text{ as before}$$

TEST FOR INDEPEDENCE

Problem 3. An experiment was conducted to determinate is Captan would be effective in controlling a plant disease. The data is recorded as number of plants which were diseased or not diseased.

		Diseased		Not diseased	Total	Ratio
Captan treated	a	4(24.94)	b	49(28.06)	53	0.5196
Not treated	c	44(23.06)	d	5(25.94)	49	0.4804
Total		48		54	102	

If Captan had no effect on the disease, the number of diseased and not diseased plants should be the same for Captan treated and not treated plants. In other words the number of observed plants in each category should be the same.Chi-square allows us to test this hypothesis. The number of calculated plants (in brackets) in each category is obtained as follows:

We first calculate the ratios for each row: row 1 = 53/102 = 0.5196, row 2 = 49/102 = 0.4804

Next we multiply the ratio by the appropiate column total: Row 1, column 1 = 0.5196 x 48 = 24.94, Row 1, column 2 = 0.5196 x 54 = 28.06, etc.

We calculate Chi-square exactly as we did before:

	O	C	(O-C)	$(O-C)^2/C$
a	4	24.94	-20.94	17.58
b	49	28.06	20.94	15.61
c	44	23.06	20.94	18.99
d	5	25.94	-20.94	16.90
Total	102	102.00		

$\chi^2 = 69.08$

$\chi^2 = 69.08$ DF = (r-1)(c-1)=1 χ^2(tabulated) 1% level = 6.63

Our Chi-square value is highly significant. Therefore there is a significant association between the treatments and the number of diseased plants. On looking at the data it is obvious that Captan is giving significant control.

A shorter method for the 2 x 2 table is available:

$$X^2 = \frac{n(ad-bc)^2}{(a+c)(b+d)(a+b)(c+d)} = \frac{102[(4x5)-(49x44)]^2}{(4+44)(49+5)(4+49)(44+5)}$$

$= 69.13$ as before

Problem 4. 230 wheat plants were classified for their reaction to Stem Rust and to Crown Rust:

	Stem Rust Reaction			
Crown Rust	Resistant	Susceptible	Total	Ratio
Resistant	66(74.87)	43(34.12)	109	0.4739
Susceptible	75(68.00)	24(30.99)	99	0.4304
Intermediate	17(15.10)	5(6.88)	22	0.0956
Total	158	72	230	

	O	C	O-C	$(O-C)^2/C$
	66	74.87	-8.87	1.0508
	75	68.00	7.00	0.7206
	17	15.10	1.90	0.2391
	43	34.12	8.88	2.3111
	24	30.99	-6.99	1.5766
	5	6.88	-1.88	0.5317
Total	230	229.96		$\chi^2 = 6.4119$

$\chi^2 = 6.4119$ DF = (r-1)(c-1) =2 $\chi^2_{.05} = 5.59$ $\chi^2_{.01} = 9.21$

Our Chi-square value is significant at the 0.05 level but not at the 0.01 level. This provides strong evidence of an association between the reaction to Stem Rust and to Crown Rust.

Problem 5. An experiment was conducted to study the effect of a number of treatments on Apple Scab. The data was recorded as the number of apple trees on which scab was either reduced or not reduced.

Treatments were:
A Single rate of Phygon
B Double rate of Phygon
C Single rate of Agristrep
D Double rate of Agristrep
E Combined Phygon and Agristrep
F Control (no treatment)

Treat.	Reduced	Notreduced	Total	Ratio	Percentage reduction
A	16 (12.14)	38 (41.86)	54	0.1759	0.2963
B	13 (10.56)	34 (36.44)	47	0.1531	0.2766
C	15 (11.69)	37 (40.32)	52	0.1694	0.2885
D	15 (11.01)	34 (37.98)	49	0.1596	0.3061
E	8 (11.91)	45 (41.09)	53	0.1726	0.1509
F	2 (11.69)	50 (40.32)	52	0.1694	0.0385
Total	69	238	307	1.0000	0.2248

	O	C	O-C	$(O-C)^2/C$
	16	12.14	3.86	1.2273
	13	10.56	2.44	0.5638
	15	11.69	3.31	0.9372
	15	11.01	3.99	1.4460
	8	11.91	-3.91	1.2836
	2	11.69	-9.69	8.0322
	38	41.86	-3.86	0.3559
	34	36.44	-2.44	0.1634
	37	40.32	-3.32	0.2734
	34	37.98	-3.98	0.4171
	45	41.08	3.92	0.3741
	50	40.32	9.68	2.3240
Total	307	307.00		$\chi^2 = 17.3980$

$\chi^2 = 17.3980$ DF = (r-1)(c-1) = 5 $\chi^2_{.01} = 15.1$

The calculated Chi-square value of 17.3980 exceeds the tabulated Chi-square value at the 0.01 level. Our Chi-square is therefore highly significant indicating significant differences between treatments.

A shorter method of calculation is available for this example:

$$\chi^2 = \frac{\text{Sum}(ap) - \bar{p}(\text{Sum } a)}{\bar{p}\bar{q}}$$

where
- a = number of trees in which scab reduced
- p = percentage reduction for each category
- $\bar{p}$ = mean percentage reduction
- $\bar{q} = 1 - \bar{p}$

$$\chi^2 = \frac{[(16 \times 0.2963) + \ldots + (2 \times 0.0385)] - 0.2248(69)}{0.2248 \times .7752}$$

$$= \frac{18.5398 - 15.5084}{0.17426} = 17.3958 \text{ as before}$$

This does not complete the range of tests or comparisons we can make in this experiment. We can partition the treatments and test the differences between individual treatments and between groups of treatments. For example, we can compare:

1. Treated vs. control (A+B+C+D+E) vs. F
2. Phygon vs. Agristrep vs. combined (A+B) vs. (C+D) vs. E
3. Between Phygon rates A vs. B
4. Between Agristrep rates C vs. D

1. <u>Treated vs. Control</u>

	Reduced		Not reduced		Total
A+B+C+D+E	a	67	b	188	255
F	c	2	d	50	52
Total		69		238	307

Because one category "c" only has two trees we will use the correction for continuity. First determinate the larger of the two crossover products, ad and bc. ad=3350, bc=376. For the larger product decrease each entry a and d by 0.5. For the smaller product increase each entry b and c by 0.5. This gives us:

		Reduced		Not reduced	Total
A+B+C+D+E	a	66.5	b	188.5	255
F	c	2.5	d	49.5	52
Total		69.0		238.0	307

We will use the short method for calculating Chi-square:

$$\chi^2 = \frac{n(ad-bc)^2}{(a+c)(b+d)(a+b)(c+d)} = \frac{307[(66.5x49.5)-(188.5x2.5)]^2}{69 \times 238 \times 255 \times 52}$$

$$\frac{307(3291.75-471.25)^2}{217{,}755{,}720.00} = 11.21$$

DF = 1 $\quad\chi^2$(tabulated) at 1% level = 6.63

Our calculated Chi-square value is significant at the 0.01 level. Therefore, treatments when combined, significantly reduced scab when compared with the control.

2.Phygon vs. Agristrep vs. Combined

	Reduced	Not Reduced	Total	Percentage Reduced
A+B	29	72	101	0.2871
C+D	30	71	101	0.2870
E	8	45	53	0.1509
Total	67	188	255	0.26274

Using the Chi-square formula $\chi^2 = \frac{\text{sum(ap)} - \bar{p}\,\text{(sum a)}}{\bar{p}\bar{q}}$

$$= \frac{[(29\text{x}0.2871) + (30\text{x}0.2970) + (8\text{x}0.1509)] - (67\text{x}0.26274)}{0.26274 \times 0.73726}$$

$$= 4.33$$

DF = (r-1)(c-1) = 2 χ^2(tabulated)at 5% level = 5.99

Our Chi-square value does not reach the level requiered for significance at the 5% level. Therefore Phygon, Agristrep and the combined treatment do not differ in their effect on apple scab.

Test number 3 and 4 can be handled in the same way. When we conduct these tests we find that the Chi-square value for the two rates of Phygon is 0.00 indicating that there is no differences between rates. The Chi-square value for the two rates of Agristrep is also 0.00 indicating no difference between the rates of Agristrep.

UNEQUAL (NON-HOMOGENEOUS) VARIANCES

and TESTS FOR HOMOGENEITY

One of the asumptions underlying the analysis of variance is that the errors are normally and independently distribuited with a common variance. In other words the variance must be homogeneous. Even trough the treatment means may differ greatly, the variance of each of the treatments should not differ greatly. As an example of situations where the variances may differ consider the following. In experiments conducted to study the effects of fungicides, insecticides or herbicides an untreated check is commonly included to measure the basic level of infestation and to provide a basis for determining the effectiveness of the treatments. The variation of the individual observations among the check plots may be much greater than the variation for the other treatments, primarily because the check will probably have a larger mean and therefore a greater base for variation. A treatment which gives almost complete and uniform control will have a small mean and very little variation from plot to plot. Still other treatments may be intermediate in effect and may give very good control in one plot and very poor control in a second plot. In this situation you can see that the variances will probabily not be homogeneous. When large departures from homogeneity occur there are differente techniques which may be used to correct for or overcome these departures. Three examples showing tests of homogeneity of variances and the recommendations made with respect to the data after the tests have been conducted are shown below.

Example 1. this example is concerned with the use of transformations. The experiment was conducted to study the effect of four herbicide treatments on the control of certain weed. A completely randomized design was used. The results, shown in Table I, are the weed counts per plot taken one month after spraying. There were tenplots of each treatment.

When we look at the means and variances for each individual treatment we see that there is a definite, positive relationship, the larger the mean, the larger the varaince. This visual relationship porvides a warning that a test of the homogeneity of variance should be consider. Two such test are shown, following Table I.

TABLE I

Weed Counts for Four Herbicide Treatments

	Treatments				
	A	B	C	D	
	20	8	50	151	
	28	10	67	125	
	36	12	90	180	
	16	16	103	149	
	25	9	90	175	
	42	12	124	178	
	25	24	72	151	
	24	10	78	152	
	31	19	70	161	
	33	10	76	118	
Total	280	130	820	1,540	2770
n	10	10	10	10	
Mean	28	13	82	154	
Sum $(x)^2$	8,376	1,926	71,118	241,086	
$(\text{Sum } x)^2/n$	7,840	1,690	67,240	237,160	
Diff. = SS	536	236	3,878	3,926	
s^2 = SS/n-1	59.55	26.22	430.89	436.22	

A. **The Barlett Test.** This is an old, well stablished test. It is a bit sensitive to non-normality, especially if the tails of the distribution are overly long, when it tends to show significance too often. For this reason it has been sugested that a significance level of 0.001 be used when the distribution seems to have long tails. This will help to prevent having yo make unnecessary transformations. To conduct the test the necessary data is laid out as shown in Table II.

TABLE II

Bartlett's Test of Homogeneity

Treatments	s^2	$\log s^2$
A	59.55	1.77488
B	26.22	1.41863
C	430.89	2.63437
D	436.22	2.63971
Total	952.88	8.46759

$s^{-2} = 952.88/4 = 238.22$ $\log s^{-2} = 2.37698$

$$M = 2.3026\ (DF)\ [k(\log s^{-2})\ -\ \text{Sum}\ \log s^2\ ]$$

where

2.3026 = log base e

DF = degrees of freedom per variance = 9

k = number of treatments = 4

$$M = 2.3026\ (9)\ [\ 4(2.37698\ -\ 8.46759]\ = 21.56$$

$$C = 1 + \frac{(k+1)}{3(DF)(k)} = 1 + \frac{5}{3(9)(4)} = 1.046$$

$$\chi^2 = M/C = 21.56/1.046 = 20.6$$

$DF = 3$ χ^2(tabulated) at 0.001 level = 16.3

Our calculated Chi-square value of 20.6 is significant at the 0.001 level. Therefore we would reject the hypothesis that the variances are homogeneous. We must re-examine our data and consider some technique, such as transforming data.

B. The Burr Foster Q-test. This test is easy to calculate and does not have the sensitivity to normality departures encoutered with the Bartlett test. It is suggested that a significance level of 0.01 be used with this test. To conduct the Burr Foster Q- test the data is laid out as shown in Table III.

TABLE III

Burr Foster Q-Test of Homonegeity

Treatment	s^2	s^4
A	59.55	3,546.20
B	26.22	687.49
C	430.89	185,666.19
D	436.22	190,287.89

$$q = \frac{(s^4a + s^4b + s^4c + s^4d)}{(s^2a + s^2b + s^2c + s^2d)^2} = \frac{380,187.77}{(952.88)^2} = 0.419$$

Degrees of freedom = v = (n-1) = 9

Look at the table "Percentile Points for Q-test, for equal degrees of freedom v and for p samples " In our experiment v = 9 and p = 4.

$q_{.01} = .0398$ $\qquad$ $q_{.001} = 0.468$

Our calculated chi-square value fall between the .01 and .001 levels and is therefore significant at the .01 level. As with the Bartlett test we reject the hypothesis of homogeneity.

Transformation of Data

Based on the evidence of the two tests, along with general evidence or experience that this kind of data usually has the size of the variance proportional to the size of the mean, it was decided that the weed count data should be transformed using the square-root transformation. The transformation of the Table I data is shown in Table IV.

TABLE IV

Transformation of Weed-Count Data by $\sqrt{y}$

	Treatments				
	A	B	C	D	
	4.47	2.83	7.07	12.29	
	5.29	3.16	8.18	11.18	
	6.00	3.46	9.49	13.42	
	4.00	4.00	10.15	12.21	
	5.00	3.00	9.49	13.23	
	6.48	3.46	11.14	13.34	
	5.00	4.90	8.48	12.29	
	4.90	3.16	8.83	12.33	
	5.57	4.36	8.37	12.69	
	5.74	3.16	8.72	10.86	
Total	52.45	35.49	89.92	123.84	301.70
n	10	10	10	10	
Mean	5.2450	3.5490	8.9920	12.3840	
Sum $(x)^2$	279.9379	129.9285	820.1142	1540.2542	
$(\text{Sum } x)^2/n$	275.1002	125.9540	808.5606	1533.6346	
Diff. = SS	4.8377	3.9745	11.5536	6.6196	
s^2 = SS/n-1	0.5375	0.4416	1.2837	0.7355	

When we look at the treatments means and variances we see that there is still a positive relationship but that it is no where near as pronounced as it was before. We will now conduct Bartlett's test for homogeneity on the transformed data.

TABLE V

Bartlett's Test of Homogeneity Using $\sqrt{y}$

Treatments	$s^2 \sqrt{y}$	$\log s^2 \sqrt{y}$
A	0.5375	-0.2696
B	0.4416	-0.3550
C	1.2837	0.1085
D	0.7355	-0.1334
Total	2.9983	-0.6495

$s^{-2} \sqrt{\bar{y}}$ = Sum $s^2 \sqrt{\bar{y}}/4$ = 2.9983/4 = 0.7496

$\log s^{-2} \sqrt{\bar{y}}$ = -0.1252

M = 2.3026 (9) [4(0.1252) - (-0.6495)] = 3.0816

C = 1.046 as before

M/C = 3.0816/1.046 = 2.9461

DF = 3 χ^2 (tabulated) at 0.05 level = 7.81

Our calculated Chi-square value of 2.9461 does not reach the 5% level and is nonsignificant. Therefore using the square-root transformation we accept the hypothesis of homogeneity of variances. We will not show it here but the Q-test also substantiates the hypothesis of homogeneity. At this point we can cayy out the analysis of variance on the transformed data. Actually we will conduct the analysis of variance on both non-transformed data (Table VI) and transformed data (Table VII).

TABLE VI

Analysis of Variance with Non-Transformed Data

Source of Variation	DF	SS	MS	F
Between treatments	3	122,107.500	40,702.500	170.86**
Error	36	8,376.000	238.222	
Total	39	130,483.500		CV=22.29%

TABLE VII

Analysis of Variance with Transformed Data

Source of Variation	DF	SS	MS	F
Between treatments	3	467.4555	155.8185	206.11**
Error	36	27.2166	0.7560	
Total	39	494.6721		CV=11.53%

Other Transformations

As mentioned , transformations are utilized to obtain a variate which, when analyzed by the analysis of variance, will conform to the basic assuptions of the analysis of variance. We have shown above the application of the square root transformation. Actually, in some cases the $\sqrt{y+0.5}$ or the $\sqrt{y+1.0}$ or even the $\sqrt{y+3/8}$ does a better job. In your specific work some other value may be still better. For "growth" type data, for example the gain in weight of animals, the most comonly used transformation is log y. The transformation to logs is also used whe the mean is proportional to the standard deviation.

In the standard deviation is proportional to the square of the mean, the reciprocal (1/y) is often used.

With percentage data or proportions dealing with the binomial, the variance is proportional to [mean(1-mean)] and the correct transformation is arc sin_p. Tables of arc sin√p are given in practically all statistic texts. As a general rule, we don't usually carry out this transformation if the data are all between √p = 0.20 to 0.80 because the variance fairly well established in this region. If the values of p or some of the values fall between 0.00 and 0.20 or between 0.80 and 1.00 the transformation of the data is carried out.

You can see from this brief discussion that many different transformations are available and used. What I have tried to do here is to introduce you to the concept of transforming data so it will conform to the basic assumptions of the analysis of variance. In experiments which you will conduct in the future, you may find that your data must be transformed. One of the above transformations may work perfectly for your data, or you may find that you will have to develop a transformation that does a better hob with the specific set of data you have obtained.

Example 2. A very important property of Chi-square is it's additive character. The sum of two chi-square's is also a chi-square for the number of degrees of freedon obtained by adding the degrees of freedom of the original chi-squares. This additive character provides a test of homogeneity from which we can determine whether it is legitimate for us to combine the data from a series of separate experiments or tests.

In this example seven genetic families were combined to obtain a composite or overall chi-square value. The danger involved in this procedure is that we may mask a kack of consistency in the ratios of the separate families. The seedlings of seven families of sweet clover were segregating for tall vs. dwarf. The results for each family and for the combined data are given in Table VIII. The hypothesis is that there will be a 3:1 segregation for tall vs. dwarf plants.

TABLE VIII

Segregation for Tall and Dwarf Plants in Sweet Clover

Family	Observed		Calculated		(O-C)	$(O-C)^2/C$
	Tall	Dwarf	Tall	Dwarf		
1	15	6	15.75	5.25	0.75	0.143
2	19	4	17.25	5.75	1.75	0.710
3	14	9	17.25	5.75	3.25	2.449
4	11	6	12.75	4.25	1.75	0.961
5	2	4	4.50	1.50	2.50	5.556
6	31	5	27.00	9.00	4.00	2.370
7	18	11	21.75	7.25	3.75	2.586
Total					X^2 =	14.775
Composite	110	45	116.25	38.75	6.25	1.344

Calculation of χ^2 in family 1 = $\chi^2 = \frac{(15-15.75)^2}{15.75} + \frac{(6-5.25)^2}{5.25} = .143$

Calculation of Composite χ^2 $\quad \chi^2 = \frac{(110-116.25)^2}{116.25} + \frac{(45-38.75)^2}{38.75} =$

$= 1.3440$

Each family yields a chi-square for 1 degree of freedom. The sum of these chi-square (14.775) is a total chi-square for 7 degrees of freedom. This total chi-square is composed of two parts:

1) A portion concerned with the grand deviation of all seven families, taken togheter, from the expected 3:1 ratio = 1.344.
2) A portion concerned with the disagreement among the families when allowance has been made for the grand deviation. This is the heterogeneity chi-square which tests the heterogeneity between the seven families.

Item	DF	χ^2	P
Deviation	1	1.344	0.30-0.20
Heterogeneity	6	13.431	0.05-0.02
Total	7	14.775	0.05-0.02

he significant p value for heterogeneity indicates porr agreement among families and suggests that more data is needed to substantiate the hypothesis that tall and dwarf plant are segregating according to a 3:1 ratio in all seven families.

Example 3. The assumption of equal variances will, in theory, obtain only when all the experiments have been conducted in the same way, with the same amount of control over environmental conditions and with experimental material of the same variability. In field, greenhouse or laboratory experiments this degree of uniformity is seldon attained. Because of random, uncontrolled error we would expect experimental error of individual experiments to differ from location to location and from year to year.

At the same time, in practice it is often desirable to combine the data from several separate experiments in order to obtain a broad overall picture of the effects of treatments at different locations and over a period of years, and alos in order to obtain an estimate of the experimental error for the pooled data. This approach is valid when all the experiments conditions, the original experiments may be considered as random samples drawn from the same population. When the variances are not homogeneous they may not be considered as coming from the same population and the pooled error is not applicable to some or all of the original experiments. When the variance are not homogeneous the general effect is that the tabulated "F" value will be too low and significance will be obtained more often than should be the case.

The following experiment conducted in Minnesota will serve as an example. Five varieties of wheat were tested at four locations over two years. There were three replicates at each location, each year. The key-out of degrees of freedom is roughly as follows:

	DF
Replicates	2
Stations	3
Years	1
Varieties	4
Stations x Years	3
Stations x Varieties	12
Years x Varieties	4
Stations x Years x Varieties	12
Pooled Error	78
Total	119

We want to apply the test for homogeneity of variance to the eight error variances which were calculated for each of the eight individual variety trials conducted at the four locations over the two years. What we want to find out is if we can combine the data from the ieght individual trials. In this example we will use Bartlett's test of Homogeneity.

The degrees of freedom for the error variance for each test equals 8 as is shown below for one station in one year:

Station A Year 1965	DF
Replicates	2
Varieties	4
Error	8
Total	14

The data is set up as shown in Table IX. This is exactly the same layout as we used in Table II

TABLE IX

Bartlett's Test of Homogeneity

Station	Year	s^2	log s^2
A	1965	5.20	0.71600
	1966	34.22	1.53428
B	1965	29.93	1.47611
	1966	9.82	0.99211
C	1965	19.30	1.28556
	1966	20.35	1.30856
D	1965	23.71	1.37493
	1966	10.16	1.00689
Total		152.69	9.69444

$$s^{-2} = \frac{\text{Sum } s^2}{k} = \frac{152.69}{8} = 19.08625 \qquad \log s^{-2} = 1.28072$$

$$M = 2.3026\ (8)\ [8(1.28072)-9.69444] = 10.15576$$

$$C = 1 + \frac{9}{3(8)(8)} = 1.0469$$

$$\chi^2 = M/C = 10.15576 / 1.0469 = 9.70$$

$$\chi^2_{.05}\text{(tabulated) for 8 DF} = 15.51$$

Our chi-square value is not significant. Therefore the eight error variances can be considered to be homogenous (samples from the same population). It is therefore statistically valid to combine the data from the eight experiments to permit a combines analysis of variance, with the calculation of one pooled error mean square fro the study of the data in all eight experiments.

If the chi-square value had been significant then the overall data must be examined. One common approach is to reject extremely high or low error variances, one at a time, until a non-significant chi-square value is obtaines. Experiments for which error variances have been rejected must not be included in the study.

EXTREME OBSERVATIONS, OUTLIERS, GROSS ERRORS

REJECTION OR EVALUATION

When experiments are conducted, it is not uncommon for errors to creep in. A measurements may be wrongly recorded, a portion of the crop harvested from a plot may be lost, a mistake may be made when applying a treatment. Such errors, especially if large, distort the mean of the treatment involved and by increasing the error variance, affect conclusions about the other treatments.

If a figure in the data looks suspicious it may be possible to go back and determine just what happened. It may even be possible to correct an incorrect figure and determine the true value. If we cannot find out what happened we will be left with a value which is much too large or much too small. Then, of course, we have to decide what to do with such an "outlier". With more complex experiments involving several factors it can be difficult to detect unusual values by eye. Instead we can calculate the "residuals" of the individual units (plots). We can calculate residuals for all the plots or just for those plots which have values which are suspiciously low or high.

In the randomized block experiment the residual, D_{ij} is calculated as follows:

$$D_{ij} = x_{ij} - \bar{x}_{i.} - \bar{x}_{.j} + \bar{x}_{..}$$

where

x_{ij} is the yield of the individual plot

$\bar{x}_{i.}$ is the mean yield of the treatment which includes that plot

$\bar{x}_{.j}$ is the mean yield of the replicate which include that plot

$\bar{x}_{..}$ is the overall mean of the entire experiment.

The results of a variety yield trial are given in Table I. The design was randomized complete block design.

TABLE I

Variety Yield Trial (cwt/acre)

Variety	Replicates 1	2	3	Totals	Means
A	1.30	3.40	6.38	11.08	3.69
B	2.26	4.79	4.09	11.14	3.71
C	2.07	3.93	6.45	12.45	4.15
D	1.74	3.51	5.56	10.81	3.60
E	2.34	4.76	5.87	12.97	4.32
F	2.34	3.82	5.23	11.39	3.80
G	1.12	3.96	5.90	10.98	3.66
H	2.30	4.31	6.49	13.10	4.37
I	2.05	4.39	6.68	13.12	4.37
Total	17.52	36.87	52.65	107.04	
Mean	1.95	4.10	5.85		3.96

TABLE II

Analysis of Variance

Source of Variation	DF	SS	MS	F
Replicate	2	68.7981	34.3990	
Variety	8	2.6400	0.3300	0.83ns
Error	16	6.3758	0.3985	

The residuals for every plot are shown in Table III. Normally you would not calculate all the residuals but only those for plots where the figures looks unusually high or low. If we look at Table I we see that one figures looks unusually low, treatment B in replicate 3. In replicate 1 treatment b is fourth from the top, in replicate 2 it is at the top but in replicate 3 it is bay far the lowest yielding variety. One other figure looks a little high, treatment F in replicate 1.

TABLE III
Residuals

Variety	Replicates 1	2	3
A	-0.38	-0.43	0.80
B	0.56	0.94	-1.51
C	-0.07	-0.36	0.41
D	0.15	-0.23	0.07
E	0.03	0.30	-0.34
F	0.55	-0.12	-0.46
G	-0.53	0.16	0.35
H	-0.06	-0.20	0.23
I	-0.31	-0.12	0.42

The residuals are calculated as follows:

Treatment A, replicate 1 $D_{A1} = 1.30 - 3.69 - 1.95 + 3.96 = -0.38$

Treatment A, replicate 2 $D_{A2} = 3.40 - 3.69 - 4.10 + 3.96 = -0.43$

If a very large residual cannot be explained then we can consider rejecting it and using the missing plot technique to replace it. Such a residual is the one for treatment B in replicate 3. Remember that such a large aberrant value is quite likely not to correct value and logic alone tells us that if it is not the correct value it would be ridiculous to retain it because our final analysis and results are incorrect.

However, and this is just as important, we don't want to reject values that are really correct or true values so in deciding whether or not to reject a value we introduce what is called "premium" or probability factor (P), which greatly reduces our chances of rejecting a values which is actually a correct one.

Anscombe and Tukey present a rule that rejects an observations whose residual has a value of, say d, if d is greater than Cs, where C is a constant to be determined from the data and s is the standard deviation of the experimental error (square root of the error mean square). For any small value of "P", say 2.5% or 5%, we can compute C.

The formula for C involves the number of error degrees of freedom and the total number of residuals "N". In this experiment f = 16 and N = 27.

The calculations involve three steps:

1.- Find the one-tailed normal deviate "z" corresponding to the probability fP/100N, where P is the premium expressed in percent.
2.- Calculate K = 1.40 x 0.85z.
3.- Calculate $C = K\left[1 - \frac{(K^2 - 2)}{4f}\right]\sqrt{f/N}$

f = 16 N = 27 P = 2.5% s = 0.6313
f/N = 0.5927 $\sqrt{f/N}$ = 0.770
$fP/100N = \frac{16x2.5}{100x27} = 0.0148$

The one-tailed normal deviate "z" is found in a table (Cumulative Normal Frequency Distribution). Look in the body of the table for the value 0.5000 - 0.0148 = 0.4852 . For the value 0.4852, z = 2.17.

K = 1.40x (0.85 x 2.17) = 2.5823
$C = 2.5823\left[1 - \frac{4.6683}{64}\right]0.770 = 2.5823 \times 0.9271 \times 0.770 = 1.8434$
Cs = 1.8434 x 0.6313 = 1.16

Let us look at Table III for variety B, replicate 3. Our residual has a value of - 1.51. This exceeds the Cs value of 1.16 and we would be justified in rejecting this observation. For variety F, replicate 1 our residual is 0.55 which does not exceeds Cs se we would not reject this observation.

The next step is to calculate a new value for Variety B, replicate 3 by means of the missing plot technique

$$X = \frac{aT + bR - S}{(a-1)(b-1)} = \frac{9(7.05) + 3(48.56) - 102.95}{(9-1)(3-1)} = 6.64$$

We replace the original figure of 4.09 with the new figure of 6.64, construct a new table with corrected totals and calculate once again the sums of square.

Correction Factor = 444.8136
$S(x)^2$ = 529.5295 - CF = 84.7159
$S(x\ rep)^2/n$ = 523.7097 - CF = 78.8961
$S(x\ treat.)^2/n$ = 448.0996 - CF = 3.2860

The error sum of squares = 84.7159 - 78.8961 - 3.2860 = 2.5338

Now remeber how to handle data when we have missing values.
We must :

1) Decrease degrees of freedom for total and error by one

2) Decrease the sum of square for varieties by:

$$\text{Correction for bias} = \frac{[R - (a-1) X]^2}{a(a-1)} =$$

$$\frac{[48.56 - (9-1)6.64]^2}{9(9-1)} = 0.2888$$

Corrected sum of squares for varieties = 3.2860 - 0.2888 = 2.9972

The analysis of variance for the corrected data is shown in Table IV.

TABLE IV

Analysis of Variance

Source of Variation	DF	SS	MS	F
Replicate	2	78.8961	39.4480	
Variety	8	2.9972	0.3746	2.22*
Error	15	2.5338	0.1689	
Total	25	84.7159	CV = 10.12%	

In this particular experiment the F value for varieties is still non-significant, but it is a much larger value. let us compare the coefficients of variation. The CV has been reduced from 15.9% to 10.12% which is very noticeable reduction in this when only one value out of 27 was changed.

A final point. Go back to Table III and calculate the sum of squares of the residuals ($-0.38^2 + 0.56^2 + -0.07^2 + ... + 0.23^2 + 0.42^2$). The total is 6.3749 which is essentially the same as the error sum of squares in the analysis of variance, Table II, which is 6.3758. In other words the residuals are the deviations of the individual plot yields from the expected values and so the sum of squares of the residuals is the measure of the random variation or error in the experiment.

CALCULATION OF RESIDUALS

This excercise will show the calculation of residuals for two-factors, three-factor and four-factor experiments. The information may be of help to you in the future because the study and evaluation of residuals is becoming more important in data analysis. The first example is a two-factor experiment consisting of three replicates and one treatment factor at three levels. The data is shown in Table I

TABLE I

Two-Factor Experiment

Treatment	Replicates 1	2	3	Total	Mean
A1	10	18	13	41	13.67
A2	19	26	23	68	22.67
A3	31	36	34	101	33.67

To put it in simple terms the calculation of residuals is as follows:

Residual =	X	- R¯	- A¯	+ G¯
	Plot yield	Rep. Mean	Treat. Mean	Grand Mean

Residual A1R1 = 10 - 20 - 13.67 + 23.33 = - 0.34

Residual A1R2 = 18 - 26.67 - 13.67 + 23.33 = 0.99

.

.

A3R3 = 34 - 23.33 - 33.67 + 23.33 = 0.33

The residuals are shown in Table II

TABLE II
Residuals for Two-Factor Experiment

Treatment	Replicates 1	2	3	
A1	-0.34	0.99	-0.67	
A2	-0.34	-0.01	0.33	Negative sum = -2.37
A3	0.66	-1.01	0.33	Positive sum = 2.31

The second example is a three factor experiment with three replicates and two treatment factors (A1 A2 A3 and B1 B2 B3). The data is shown in Table III.

TABLE III
Three-Factor Experiment

Treatments A	B	Replicates 1	2	3	Total	Mean
A1	B1	10	18	13	41	13.67
	B2	19	26	23	68	22.67
	B3	31	36	34	101	33.67
Total		60	80	70	210	
Mean		20.00	26.67	23.33		23.33
A2	B1	18	26	22	66	22.00
	B2	32	32	36	100	33.33
	B3	40	51	42	133	44.33
Total		90	109	100	299	
Mean		30.00	36.33	33.33		33.22
A3	B1	24	29	32	95	31.67
	B2	41	44	43	128	42.67
	B3	48	59	54	161	53.67
Total		113	142	129	384	
Mean		37.67	47.33	43.00		42.67
Total B1		52	83	67	202	
Mean		17.33	27.67	22.33		22.44
Total B2		92	102	102	296	
Mean		30.67	34.00	34.00		32.89
Total B3		119	146	130	395	
Mean		39.67	48.67	43.33		43.89
Total		263	331	299	893	
Mean		29.22	36.78	33.22		33.07

The calculations of residuals is as follows:

$$\text{Residual} = X + \bar{R} + \bar{A} + \bar{B} - \bar{R}\bar{A} - \bar{R}\bar{B} - \bar{A}\bar{B} - \bar{G}$$

X	$\bar{R}$	$\bar{A}$	$\bar{B}$	$\bar{R}\bar{A}$	$\bar{R}\bar{B}$	$\bar{A}\bar{B}$	$\bar{G}$
Plot Yield	Rep. Mean	Treat. A Mean	Treat. B Mean	R x A Mean	R x B Mean	A x B Mean	Grand Mean

Residual A1B1R1 = 10+ 29.22 + 23.33 + 22.44 - 20.00 - 17.33 - 13.67 - 33.07 =
= 0.92

Residual A1B1R2=18.0+36.78+ 23.33 + 22.44- 26.67- 27.67- 13.67- 33.07=
= -0.53

Residual A3B3R2=59.0+ 36.78 + 42.67 + 43.89 - 47.33 - 48.67- 53.67 - 33.07 =
= -0.40

Residual A3B3R3=54.0+33.22 + 42.67 + 43.89 - 43.00 - 43.33 - 53.67- 33.07 =
= 0.71

The residuals for the entire experiment are shown in Table IV

TABLE IV
Residuals for Three-factor experiment

Treatment		Replicates 1	2	3	
A1	B1	0.92	-0.53	-0.41	
	B2	-1.97	2.59	-0.63	
	B3	1.03	-2.08	1.04	
A2	B1	0.48	-0.63	0.15	Negative sum
	B2	0.26	-1.84	1.60	-14.36
	B3	-0.74	2.49	-1.73	Positive sum
					14.38
A3	B1	-1.41	1.15	0.26	
	B2	1.70	-0.73	-0.96	
	B3	-0.30	-0.40	0.71	

Table of Means

	Rep 1 C1	Rep 1 C2	Rep 1 T	Rep 2 C1	Rep 2 C2	Rep 2 T
A1B1			16.00			23.50
B2			23.00			30.00
B3			36.50			42.00
Total	20.00	30.33	25.17	26.67	37.00	31.83
A2B1			22.00			31.50
B2			38.00			37.00
B3			44.00			57.00
Total	30.00	39.33	34.67	36.33	47.33	41.83
A3B1			27.50			42.50
B2			47.50			50.00
B3			52.50			65.50
Total	37.67	47.33	42.50	47.33	58.00	52.67
B1	17.33	26.33	21.83	27.67	37.33	32.50
B2	30.67	41.67	36.17	34.00	44.00	39.00
B3	39.67	49.00	44.33	48.67	61.00	54.83
Total	29.22	39.00	34.11	36.78	47.44	42.11

	Rep 3 C1	Rep 3 C2	Rep 3 T	Total C1	Total C2	Total T
A1B1			17.50	13.67	24.33	19.00
B2			28.00	22.67	31.33	27.00
B3			38.00	33.67	44.00	38.83
Total	23.33	32.33	27.83	23.33	33.22	28.28
A2B1			26.00	22.00	31.00	26.50
B2			42.00	33.33	44.67	39.00
B3			47.50	44.33	54.67	49.50
Total	33.33	43.67	38.50	33.22	43.44	38.33
A3B1			37.50	31.67	40.00	35.83
B2			46.50	42.67	53.33	48.00
B3			58.00	53.67	63.67	58.67
Total	43.00	51.67	47.33	42.67	52.33	47.50
B1	22.33	31.67	27.00	22.44	31.78	27.11
B2	34.00	43.67	38.83	32.89	43.11	38.00
B3	43.33	52.33	47.83	43.89	54.11	49.00
Total	33.22	42.56	37.89	33.07	43.00	38.04

ANALYSIS OF COVARIANCE

The analysis of covariance makes use of the concepts of both the analysis of variance and of regression. Covariance is basically concerned with the relationships between two or more measured variables. It is very often used where no exact control has been exercised over those measurable variables which are regarded as independent but in other cases both the dependent and independent variables are controlled.

Uses of Covariance Analysis

1. To assist in the interpretation of data. There may be for example a question as to whether treatments are having an effect on both the dependent and independent variables. Covariance will aid in aswering such questions and may aid in determining how such effects occur. Also if an effect on both the dependent and independent variables are shown to occur this information will greatly influence the interpretation of the results.

2. To control error and increase precision. One of the most important applications of covariance is in the control of random, yet assignable errors that arise throughout the experiment. For example, in an experiment plant stand may vary, at random, from plot to plot. In so far as variations in stand affects yield the varying stands in our experiment add to the random experimental error. Correcting yield on the basis that yield is directly proportional to stand is often done, but this is not correct because yield is rarely directly proportional to stand and such a correction tends to overstimate the yield of plots with missing plants. Correction on the basis of the exact relationship between yield and stand is, however, statistically valid and the method of making such a correction is a natural development of the covariance technique.

3. To adjust treatment means of the dependent variable (yield) for differences in the dependent variable (stand) That is, we adjust the treatment means for yield by regression so that they become estimates of what the yield means would have been if the treatments had all had an equal stand. Other examples of where treatment means could be adjusted are:

   Original weight vs. final weight of animals in a feeding trial.
   Amount of food consumed by animals in a feeding trial.
   Number of insects per plot before soil insecticides were applied.

Partitioning Covariance

If you think back to the analysis of variance you will remember that for a single variable in an experiment, say yield, the variation is heterogeneous and that this variation can be sorted into recognizable and assignable causes or components. In a randomized complete block design the components are blocks, treatments and error. The same thing is true for the correlated variability of covation of two variables, say stand and yield, and the mechanism for sorting out the covariance effects is known as the analysis of covariance.

For example, suppose that the two variables are the stand and yield of several varieties in a blocked trial. The total covariance for stand and yield is made up in part by the covariance of the means of the varieties and in part by the covariance of the residual or error variation. In other words the components may be taken as exactly equivalent to those according to which the data may be classified for an anlysis of variance for either variable.

The total variation of the above example (perhaps it will help if we think in our minds of the correlation coefficient of the total) is not capable of definite interpretation because it is made up of heterogeneous covariance effects and a study of the total covariance (or total correlation coefficient) cannot give us any clear cut answers. The coefficients calculated for each component; variaties, blocks and error are however capable of definite, clear cut interpretation.

In order to demostrate the covariance technique we will start by representing the analysis of covariance algebraically for a randomized complete block design, as shown in Table I.

TABLE I
Analysis of Covariance for the Randomized Complete Block Design

Source of Variation	DF	SS_X	SS_{XY}	SS_Y	$b_{y.x}$
Replicate	r	R_{XX}	R_{XY}	R_{YY}	$b_R = R_{XY}/ R_{XX}$
Treatment	t	T_{XX}	T_{XY}	T_{YY}	$b_T = T_{XY}/ T_{XX}$
Error	e	E_{XX}	E_{XY}	E_{YY}	$b_E = E_{XY}/ E_{XX}$
T + E	s	S_{XX}	S_{XY}	S_{YY}	$b_S = S_{XY}/ S_{XX}$

...continued

Source of variation	$b_{yx} \cdot SS_{XY}$	$SS_{Y'}$	new DF
Replicate	$b_R \cdot R_{XY}$	$R_{YY} - b_R \cdot R_{XY}$	r-1
Treatment	$b_T \cdot T_{XY}$	$T_{YY} - b_T \cdot T_{XY}$	t-1
Error	$b_E \cdot E_{XY}$	$E_{YY} - b_E \cdot E_{XY}$	e-1
T + E	$b_S \cdot S_{XY}$	$S_{YY} - b_S \cdot S_{XY}$	s-1

In the column heading $b_{y.x}$ is the regression coefficient for the regression of y on y and $SS_{y'}$ is the sum of squares for y adjusted by the regression coefficient in the same row. The second step is shown in Table II.

TABLE II

Source of Variation	DF	$SS_{Y'}$	MS
T + E	s-1	$S_{YY} - b_S \cdot S_{XY}$	
Error	e-1	$E_{YY} - b_E \cdot E_{XY}$	V1
Treatment 1	t	$T_{YY} + b_E \cdot E_{XY} - b_S \cdot S_{XY}$	V2
Treatment 2	t-1	$T_{YY} - b_T \cdot T_{XY}$	V3
$b_T - b_E$	1	$b_T \cdot T_{XY} + b_E \cdot E_{XY} - b_S \cdot S_{XY}$	V4

The first sum of squares for treatments is obtained by substraction and since it has not been corrected by the regression coefficient for treatments it is still represented by t defrees of freedom. The second sum of squares for treatments (Treatment 2) is obtained from Table I. It is represented by t-1 degrees of freedom since it has been adjusted by the treatment regression coefficient.

The test of significance of the treatment differences after adjustment for the rror regression of Y on X involves comparison of the variances V_1 and V_2. Why do we arrive at adjusted treatment sum of squares by substraction. Because if b_T has some effect different than b_E that affect should be expressed in the treatment sum of squares obtained by substraction. The fact that v2 may contain a significant effect due to differences between b_T and b_E does not variate the meaning of the test as such an effect is obviously due to some factor characteristic of the treatment.

A further test may be applied to V_3 and V_4 and by comparison of the significance of V_3 and V_4 a complete picture of treatment effects may be obtained. The value of such an analysis, if stand does have a significant effect on yield is that the error variance and treatment variance will be reduced proportionately, with a consequent increase in the significance of the treatment differences. If the analysis of variance of the unadjusted yields shows significant differences between treatments and the analysis of variance of the adjusted yield does not this simply means that the original differences were due to stand and not to true differences in yield.

When $b_T = b_E$ the sum of squares will, of course, be zero and a test of significance of the corresponding variance, V_4, is a test of the significance of the difference between the treatment and error regression coefficients. The treatment regression coefficient, b_T, might be higher than b_E and will contribute to the significance of V_2, but b_E represents the regression of yield on stand within treatments and may be taken as the true measure of the effect of stand on yield.

Analysis of Covariance

We will now go through the steps involved in the analysis of covariance using actual data from a popato fertilizer trial. The two variables were stand (X) and yield (Y). Stand is the independent variable, yield is the dependent variable. The design was randomized complete block design with five replicates and twelve treatments. The stand per plot and the yield per plot are shown in Table III. The sum of squares for stand and for yield and the sum of products (XY) are shown below Table III.

Before conducting the analysis of covariance it is usual to conduct the analysis of variance on the original yield data to see if there are significant differences between treatments. We have done this in Table IV.

TABLE IV
Original Analysis of Variance on Yield

Source of Variation	DF	SS	MS	F
Replicates	4	854.2333	213.5583	
Treatments	11	63,609.2500	5,782.6591	15.38**
Error	44	16,540.1667	375.9129	
Total	59	81,003.6500	CV = 11.98 %	

There is a highly significant differences between treatments. We could end our analysis at this point and simply determine wich treatments differ from which other treatments. However, we notice that there is a marked difference in stand from plot to plot, from 17 to 37 plants per plot. From past experience we know that the stand usually has a noticeable effect on yield. It is therefore possible that the differences in yield are due to differences in stand. On the other hand, if the differences in stand are completely random, that is if they are affecting yield per plot but are not influencing treatment differences, then by taking stand differences into account we can perhaps reduce the experimental error and increase the precision of the experiment. We decide therefore to conduct the analysis of covariance.

TABLE III

Stand (X) and Yield (Y) per Plot of Potatoes

Treatments		Rep. 1		Rep. 2		Rep. 3		Rep. 4		Rep. 5		Total	
N	P	X	Y	X	Y	X	Y	X	Y	X	Y	X	Y
1	1	24	89	27	54	27	77	32	67	21	108	131	395
1	2	27	114	35	103	29	133	30	151	27	131	148	632
1	3	28	160	25	186	31	142	29	160	32	153	145	801
2	1	28	114	29	136	27	132	28	147	26	140	138	669
2	2	24	206	29	147	25	150	33	165	30	169	144	837
2	3	26	185	33	151	27	183	30	188	31	188	147	895
3	1	17	195	27	152	31	160	28	174	29	150	132	831
3	2	29	164	29	192	28	192	31	182	30	166	147	896
3	3	32	164	23	208	27	184	31	186	23	214	136	956
4	1	30	194	32	184	33	165	31	181	35	159	161	883
4	2	33	160	30	163	22	226	30	195	25	236	138	1021
Total		326	1946	352	1861	338	1949	370	1954	338	2001	1724	9711

Stand

$CF = (1724)^2 / 60 = 49{,}536.26666667$

$Sum\ (X)^2 = (24^2 + 27^2 + ... + 25^2) - CF = 747.73333333$

$Sum\ (X\ rep)^2/n = (326^2 + ... + 338^2)/12 - CF = 94.40000000$

$Sum\ (X\ Treat.)^2/n = (131^2 + ... + 138^2)/5 - CF = 188.13333333$

Yield

$CF = (9711)^2/60 = 1{,}571{,}725.35000000$

$Sum\ (Y)^2 = (89^2 + 114^2 + ... + 236^2) - CF = 81{,}003.65000000$

$Sum\ (Y\ rep.)^2/n = (1946^2 + ... + 2001^2)/12 - CF = 854.23333333$

$Sum\ (Y\ treat.)^2/n = (395^2 + ... + 1021^2)/5 - CF = 63{,}609.25000000$

Cross Products (stand x yield)

$CF = (1724 \times 9711)/60 = 279{,}029.40000000$

$Sum\ (XY) = (24 \times 89) + (27 \times 114) + ... + (25 \times 236) - CF = -1{,}181.40000000$

$Sum\ (XY\ rep)/n = [(326 \times 1946) + ... + (338 \times 2001)] / 12 - CF = -67.06666667$

$Sum\ (XY\ treat)/n = [(131 \times 395) + ... + (138 \times 1021)]/5 - CF = +1{,}138.00000000$

TABLE V
Analysis of Covariance

Column Source of Variation	1 DF	2 SSX	3 Sum XY	4 SSY
Replicates	4	94.40000000	- 67.06666667	854.2333333333
Treatments	11	188.13333333	+1,138.00000000	63,609.2500000000
Error	44	465.20000000	-2,252.33333333	16,540.1666666667
T + E	55	653.33333333	-1,114.33333333	80,149.4166666667
Total	59	747.73333333	-1,181.40000000	81,003.6500000000

...continued

Source of Variation	5 = 3/2 $b_{y.x}$	6 = 5x3 $b_{y.x}SXY$	7 = 4-6 Adjusted SS Yield SSY'	8 New DF
Replicates	-0.7104	47.6442	806.5891	3
Treatments	+6.0439	6,883.6482	56,725.6018	10
Error	-4.8416	10,904.8969	5,635.2698	43
T + E	-1.7056	1,900.6069	78,248.8099	54
Total				58

...continued

Source of Variation	9 $\sqrt{SSX}$	10 $\sqrt{SSY}$	11 = 9x10 $\sqrt{SSX \cdot SSY}$	12 = 3/11 r
Replicates	9.7160	29.2273	283.9724	-0.236
Treatments	13.7162	252.2087	3,459.3450	+0.329
Error	21.5685	138.6086	2,773.8946	-0.812**
T + E	25.5604	283.1067	7,236.3205	-0.154
Total	27.3447	284.6114	7,782.6134	-0.152

The first step in the analysis of covariance is to set up Table V. The explanation of the columns in Table V is as follows:

1 = Degrees of fredom from the original sum of squares.
2 = Sum of squares for stand (X).
3 = Sum of products for stand and yield (XY).
4 = Sum of squares for yield (Y).
5 = The regression coefficient of yield on stand ($b_{x.y}$)
6 = The adjustment factor. The value which shows the degree to which the sum of squares for yield is influenced by stand.
7 = The adjusted sum of squares for yield.
8= The new degrees of freedom for the adjusted sum of squares.
9 = The square root of the sum of squares for stand.
10 = The square root of the sum of squares for yield.
11 = The divisor for calculating the correlation coefficient.
12 = The correlation coefficient.

In many examples of covariance analysis in the text books only columns 1 to 8 are shown. Since we are mainly or enterily interested in regression and a measure of the dependence of yield on stand this is all that it is really necessary to calculate. However it is often true that correlation is easier to undestand and interpret that regression. Therefore I have gone through the steps involved in the calculation of the correlation coefficients for each component and these are shown in columns 9 to 12. Because they are somewhat easier to undestand we will first look at the correlation coefficients.

The only significant correlation is that for the error component ($r = -0.812^{**}$). This significant correlation coefficient indicates that stand is affecting yield regardless of what treatment is applied or regardless of whether any treatment is applied. That is, there is a basic, underlying relationship between stand and yield. This relationship causes us to question our confidence in the significance of differences in yield are partly or largely due to differences in stand. The correlation coefficient for the treatment component ($r = +0.329$) is not significant. From this we conclude that there is not a strong or meaningful relationship between the stand and yield of individual treatments. In other words, although there are differences in stand between treatments, these differences have happened more or less by chance.

Let us go a step further and examine individual correlation coefficients in more detail. We will assume the following results:

Source of Variation	DF	r
Replicates	3	-0.135
Treatments	10	-0.782**
Error	43	-0.713**
Total	58	-0.719**

The correlation coefficient for the total ($r = -0.719$) is highly significant. As we mentiones before this correlation is not capable of definite interpretation because it is made up of a number of distinct and heterogeneous effects.

The correlation coefficient for the treatment component is highly significant. This means that certain of the treatments tended to have better stands over the five replicates than other treatments and that these stand differences were affecting yield significantly. This significant correlation explains part of and perhaps a large part of the reason for the difference in yield. The significant correlation coefficient on the error line indicates that stand does have a true or basic effect on yield and this significant correlation further weakens our confidence in the significant F value we obtained above.

Accordingly we procced to compute the regression coefficients and then conduct a test of significance on yield after adjustment for stand. The preliminary calculations have been shown in Table V and the final calculations are shown in Table VI>

The regression coefficient values on each line of Table V are obtained from $b_{x.y} = SXY/SX^2$. For example the error regression coefficient equals :

-2,252.3333/465.2000 = -4.8416

That part of the sum of squares for yield (SSY) that is due to regression is equal to $b_{yx}SXY$. Taking the error line as an example again :

-4.8416 x -2,252.3333 = 10,904.8969

The sum of squares of the errors of estimate (deviation from regression) is the difference SSY - (SS due to regression). For the error line this is:

-16,540.1667 - 10,904.8969 = 5,635.2698

There is a short cut method for obtaining the adjusted sum of squares for yield. We will use the short cut method later so I will outline it here. If we look at Table V, the adjusted sum of squares for yield can be computed as follows:

$SSY' = SSY - SXY^2/SSX$ for the error line this is

$= 16{,}540.1667 - (-2{,}252.3333)^2 / 465.20000 = 5{,}635.2698$ as before

TABLE VI

Testing Adjusted Sums of Squares for Yield

Source of Variation	DF	Adjusted SS Yield	MS	F
Treat + Error	54	78,248.8098		
Error	43	5,635.2698	131.0528	
Treat 1 (diff)	11	72,613.5400	6,601.2309	50.37**
Treat 2	10	56,725.6018	5,672.5602	43.28**
b_T-b_E	1	15,887.9382	15,887.9382	121.23**

First we will explain the degrees of freedom listed in Table VI. The DF for treatment + error are 11 + 44 - 1 = 54 because one degree of freedom belongs to regression and 54 is the correct DF finding a new mean square based on the error of estimate (deviation from regression). For the same reason the DF for the adjusted error are 44 - 1 = 43. The degree of freedom for Treatment 1 are left at eleven because the corresponding sum of squares (obtained by substraction) has not been adjusted by it's own regression coefficient. The degrees of freedom for Treatment 2 are 11 -1 = 10 because the treatment sum of squares has been adjusted by the treatment regression coefficient. Finally the sum of squares for b_T - b_E has 1 DF because both the degrees of freedom and the sum of squares are obtained by substracting Treatment 2 from Treatment 1.

The sum of squares for the errors of estimate for Tratment 1 is simply 78,248.8098 - 5,635.2698 = 72,613.5400. Tis is our new sum of squares for treatments, with error control applied by adjusting with the error regression of yield on stand. Using the new mean squares we find that the "F" test for treatments after adjustment is 6,601.2309/131.0528 = 50.37**. This is much larger than before, indicating that part of the difference between treatment was obscured by differences in stand. If the new "F" had been smaller it wiuld indicate that differences in stand were actually the cause of part of the differences between treatment yields.

The test of treatment 2, 5, 672.5602/131.0528 = 43.28**, is also highly significant. This is a test of the differences between treatment yields if all stand were equalized. If this test were not significant it would indicate that the differences in yield between treatments were due solely or largely to differences in stand.

The next step, generally carried out in covariance analysis, is to test the significance of the error regression and the treatment regression coefficient.

Error $b_{x.y}$ $\quad s_e = \sqrt{\frac{\text{Sum}(y-Y)^2}{n-2}} = \sqrt{\frac{5,635.2698}{43}} = 11.4478$

$$s_b = \frac{s_e}{\sqrt{\text{Sum }(x-x^-)^2}} = \frac{11.4478}{\sqrt{465.2000}} = 0.5308$$

$$t = \frac{b_{x.y}}{s_b} = \frac{-4.8416}{0.5308} = 9.12^{**}$$

Treatment $b_{y.x}$ $\quad s_e = \sqrt{\frac{56,725.6018}{10}} = 73.3164$

$$s_b = \frac{75.3164}{\sqrt{188.1333}} = 5.4910$$

$$t = \frac{+6.0489}{5.4910} = 1.10 \text{ ns}$$

The error regression coefficient is highly significant but the treatment regression coefficient is not significant. These results agree with those obtained from the correlation coefficients. This shows us that stand is affecting yield randomly from plot to plot but that stand does not differ in a predictable manner from treatment to treatment and that stand is not being significantly influenced by the treatments.

Adjusted Treatment Means

By use of the regression equation $Y = \bar{y} - b(x-\bar{x})$ we can calculate the expected values of the treatment means for yield after adjustment with the error regression coefficient. In the equation:

Y = the expected or adjusted treatment mean.
$\bar{y}$ = original nean yield for the specific treatment.
b = error regression coefficient
x = mean stand for the specific treatment.
$\bar{x}$ = overall mean stand for the entire experiment.

The original and corrected mean yields for each treatment are shown in Table VII.

TABLE VII

Treatment Means Corrected on
Basis of Error Regression Coefficient

Treatment		Stand	Original Yield	Corrected Yield	Corrected - Original
1	1	26.2	79.0	66.8	-12.2
1	2	29.6	126.4	130.6	4.2
1	3	29.0	160.2	161.5	1.3
2	1	27.6	133.8	128.3	- 5.5
2	2	28.8	167.4	167.7	0.3
2	3	29.4	179.0	182.2	3.2
3	1	26.4	166.2	154.9	-11.3
3	2	29.4	179.2	182.4	3.2
3	3	27.2	191.1	183.8	- 7.4
4	1	32.2	176.6	193.4	16.8
4	2	31.4	179.0	191.9	12.9
4	3	27.6	204.2	198.7	- 5.5
Mean		28.73	161.85		

The corrected yields are obtained as follows:

$$Y = \bar{y} - b(x - \bar{x})$$

$$= \bar{y} - [-4.8416(x - 28.73)]$$

$$= \bar{y} + 4.8416x - 139.10$$

For example Treatment 1 1 = 79.0 + 126.85 - 139.10 = 66.8
Treatment 1 2 = 164.4 + 143.31 - 139.10 = 130.6

In Table VII you will notice that the correction rang from -12.2 to +16.8 cwt. wich sums to 29.0. This is fairly large overall range of corrections yet the ranking of the individual treatments, except for treatments 4-1 and 4-2, changes very little.

Predicted Regression Line

We can also use the regression equation to lay out the predicted regression line.

$$Y = \bar{y} + b(x - \bar{x})$$

$$= 161.85 + [-4.8416(x - 28.73)]$$

$$= 300.95 - 4.8416x$$

In this case $\bar{y}$ equals the grand overall mean yield for the entire experiment. To fit the predicted regression line the simplest proccedure is to take two extreme values for stand, say 20 and 40, and insert these values in the equation as follows:

Stand = 20 $Y = 300.95 - 4.8416(20) = 204.12$
Stand = 40 $Y = 300.95 - 4.8416(40) = 107.30$

The predicted line is shown in Figure 1.

Increase in Prediction due to Covariance

Before the analysis of covariance was conducted the error mean square for yield was 375.9129. A comparable figure when covariance is used is the effective residual mean square.

$$\text{Effective Residual MS} = s_{y \cdot x}^2 \quad 1 + t_{xx} / e_{xx} \quad \text{where}$$

$s_{y \cdot x}^2$ = new error mean square = 131.0528

t_{xx} = treatment mean square for stand =188.1333/11= 17.1030

e_{xx} = error sum of squares for stand = 465.2000

$$\text{Effective Residual MS} = 131.0528 \quad 1 + \frac{17.1030}{465.2000} = 135.8710$$

The term in brackets represents the average contribution to error from stand. In other words, the contribution from random errors in the dependent or adjustment factor. The increase in precision obtained through covariance relative to that without covariance is estimated by 375.9129/135.8710 = 2.77. This is a considerable increase in presicion. Using the covariance technique we have almost tripled the precision of the experiment.

QUADRATIC REGRESSION

In many, probably most agricultural experiments, the relationship between the independent and dependent variable is truly linear or assumed to be linear and a linear relationship is often a reasonably good approximation for a non-linear raltion provided the values of the independent variable do not cover too wide a range. However, at times a quadratic regression will fit the data better than a linear regression and fitting a quadratic regression will give us better cibtrol of error and will permit us to better interpret the results of the experiments.

The quadratic regression equation is as follows:

$Y = y^{-} \pm b(x - x^{-}) \pm c(x^2 - x^{-2})$ where

b = the linear regression coefficient

c = the quadratic regression coefficient

x = stand

x^2 = stand squared

In Table VIII we have shown each individual stand value squared. Below the Table we have shown the sum of squares for stand squared, the sum of products for stand x stand squared (XX^2) and the sum of products for yield x.

The linear regression coefficient b and the quadratic regression coefficient c are calculated by means of simultaneous equation as follows:

$$b.S(x - x^{-}) + c.S(x - x^{-})(x^2 - x^{-2}) = S(x - x^{-})(y - y^{-})$$
$$b.S(x - x^{-})(x^2 - x^{-2}) + c.S(x^2 - x^{-2}) = S(x^2 - x^{-2})(y - y^{-})$$

We use the values from the error component and obtain :

$$b(465.200000) + c(25,657.066667) = -2,252.333333$$
$$b(25,657.0667) + c(1,437,146.4000) = -125,091.6333$$

Solution of the above simultaneous equation will lead to the best possible estimates of b and c. To obtain b we multiply the first line by 1,437.146.4000 and the bottom line by 25,657.00667. We then substract the bottom line from the top line. This gives us:

b(668,560,505.2800) + c(36,872,961,042.4648) = -3,236,932,693.6951
b(658,285,071.6482) + c(36,872,961,042.4686) = -3,209,484,379.19
b(10,275,433.6318) 0 = - 27,448,314.5051

b = -27,448,314.5051 / 10,275,433.6318 = -2.6712

TABLE VIII
Stand Squared

Treatment		1	2	3	4	5	Total
1	1	576	729	729	1024	441	3499
1	2	729	1225	841	900	729	4424
1	3	784	625	961	841	1024	4235
2	1	784	841	729	784	676	3814
2	2	576	841	784	1089	900	4190
2	3	676	1089	729	900	961	4355
3	1	289	729	961	784	841	3604
3	2	841	841	784	961	900	4327
3	3	1024	529	729	961	529	3772
4	1	900	1024	1089	961	1225	5199
4	2	1089	900	784	1369	841	4983
4	3	784	1089	484	900	625	3882
Total		9052	10462	9604	11474	9692	50284

Sum of Squares for Stand Squared

CF = $(50284)^2/60$ = 42,141,344.2667

Sum$(x^2)^2$ = $(576^2 + 729^2 +...+ 625^2)$ = 44,468,164.0000 - CF = = 2,326,819.7333

Sum$(x^2$ rep$)^2/n$ = $(9052^2 +...+ 9692^2)/12$ = 42,434,708.6667 - CF = 293,364.4000

Sum$(x^2$ treat.$)^2/n$ = $(3499^2 +...+ 3882^2)/5$ = 42,737,653.200 - CF = 596,308.9333

Sum of Products for Stand x Stand Squared

CF = (1724 x 50284)/60 = 1,444,826.9333

Sum(xx^2) = (24 x 576) +...+ (25 x 625) = 1,486,304.0000 -CF = 41,477.0667

Sum $(xx^2$ rep$)/n$ = [(326 x 9052) +...+ (338 x 9692)]/12 = 1,450,083.6667 - CF = 5,257.3333

Sum$(xx^2$ treat$)/n$ = [(131 x 3499) +...+ (138 x 3882)]/5 = 1,455,389.600 - CF = 10,562.6667

Sum of Products for Yield x Stand Squared

CF = (9711 x 50284) / 60 = 8,138,465.4000

Sum(yx^2) = (89 x 576 +...+ (236 x 625) = 8,074.918.0000 - CF = -63,547.4000

Sum$(yx^2$ rep$)/n$ = [(1946 x 9052) +...+ (2002 x 9692)]/12 = 8,134,754.8333 - CF = -3,710.5667

Sum$(yx^2$ trat$)/n$ = [(395 x 3499) +...+ (1021 x 3882)]/5 = 8,203,720.2000 - CF = +65,254.8000

Source of Variation	SS(stand)^2	Sum(stand)(stand)^2	Sum(yield)(stand)^2
Replicate	293,364.4000	5,257.3333	- 3,710.5667
Treatment	596,308.9333	10,562.6667	+ 65,254.8000
Error	1,437,146.4000	25,657.0667	-125,091.6333
T + E	2,033,455.3333	36,219.7333	- 59,836.8333
Total	2,326,819.7333	41,219.7333	- 63,547.4000

Substracting b in the first line above we obtain :

$$b(465.2000) + c(25,657.0667) = -2,252.3333$$
$$-2.6712(465.2000) + c(25,657.0667) = -2,252.333$$
$$25,657.0667c = -1,009.6911$$
$$c = -0.039353$$

The variations "due to the quadratic regression" are calculated as follows:

$$b.S(x - x^{-})(y - y^{-}) + c.S(x^2 - x^{-2})(y - y^{-}) =$$
$$-2.6172(-2,252.3333) + -0.039353(-125,091.6333)$$
$$6,016.4327 + 4,922.7310 = 10,939.1637$$

The test of significance of the quadratic regression coefficient is shown in Table IX.

TABLE IX

Test of Quadratic Regression Coefficient

Source of Variation	DF	SS	MS	F
Quadratic Regression	2	10,939.1637	5,469.5818	41.01**
Deviation from Reg.	42	5,601.0030	133.3572	
Error	44	16,540.1667		

There are now two degree freedom due to regression since two constants are used in the equation. The error term (16,540.1667) is obtained from Table V, column 4. The Deviations from Regression term is obtained by substraction.

The significant "F" value indicates that the quadratic regression coefficient is significant and therefore accounting for a significant portion of the variation in yield. However we also want to see if the quadratic regression describes the relationship between stand and yield better than the linear regression. To do this we substract the sum of squares due to linear regression from the sum of squares due to quadratic regression and compare the difference with the deviation from quadratic regression. Such a test will establish whether the curved regression line (quadratic) fits the data better than a straight line (linear). The test is shown in Table X.

TABLE X

Test of Increased Presicion of Quadratic over Linear Regression

Source of Variation	DF	SS	MS	F
Quadratic- Linear Reg.	1	-34.2668	-34.2668	0.26
Deviation from Quadratic	42	5,601.0030	133.3572	
Deviation from Linear	43	5,635.2698		

The deviations from linear regression term is obtained from Table V column 7. The deviations from quadratic regression term is obtained from Table IX and the Quadratic minus Linear Regression term is obtained by substraction.

The F value is obviously non-significant and so we conclude that the quadratic regression is no better than the linear relationship and really adds nothing of value to the study.

We have already tested the significance of the linear regression by means of the "t" test but let us do it again by means of the F test. The test is shown in Table XI.

TABLE XI

Test of Linear Regression Coefficient

Source of Variation	DF	SS	MS	F
Linear Regression	1	10,904.8969	10,904.8969	83.21**
Deviation from Reg.	43	5,635.2698	131.0528	
Error	44	16,540.1667		

The quadratic regression did not show any improvement over linear regression and so in this experiment we would make no further use of the results of the quadratic study but would correct our treatment mean yields using the linear regression coefficient as we did in Table VII. However I will go through the steps involved in adjusting treatment yields by means of the quadratic regression. The equation fro correcting the treatment means is:

$$Y = y^{-} - b(x - x^{-}) - c(x^{2} - x^{-2})$$

where

Y = the expected or adjusted treatment mean.

y^{-} = the original mean yield for the specific treatment.

b = linear regression coefficient = -2.6712

c = quadratic regression coefficient = -0.03935

x = mean stand for the specific treatment.

x^{-} = overall mean stand for the entire experiment = 28.73

x^{2}= mean stand squared for the specific treatment.

x^{-2}= overall mean stand squared for the entire experiment = 838.07

The original treatment mean yields and the mean yields corrected by both the linear and the quadratic regression are shown in Table XII. If you look at the two corrected columns you will see that there is practically no difference between them. This indicates that linear regression is as good as quadratic regression and that quadratic regression provides really no additional or meaningful correction.

TABLE XII
Treatment Yields Corrected by Linear and by Quadratic Regression

Treatment		Stand	Stand Squared	Original Yield	Corrected Yield Linear Reg.	Quad. Reg.
1	1	26.2	699.8	79.0	66.8	66.8
1	2	29.6	884.8	126.4	130.6	130.6
1	3	29.0	847.0	160.2	161.5	161.3
2	1	27.6	762.8	133.8	128.3	127.9
2	2	28.8	838.0	167.4	167.7	167.6
2	3	29.4	871.0	179.0	182.2	182.1
3	1	26.4	720.8	166.2	154.9	155.4
3	2	29.4	865.4	179.2	182.4	182.1
3	3	27.2	754.4	191.2	183.8	183.8
4	1	32.2	1039.8	176.6	193.4	193.8
4	2	31.4	996.6	179.0	191.9	192.4
4	3	27.6	776.4	204.2	198.7	198.8
Mean		28.73	883.07	161.85		

The corrected yields, using the quadratic regression are obtained as follows:

$$Y = \bar{y} - b(x - \bar{x}) - c(x^2 - \bar{x}^2)$$

$$= \bar{y} - [-2.6712(x-28.73)] - [-0.03935(x^2-838.07)]$$

$$= \bar{y} + 2.6712x - 76.74 + 0.03935x^2 - 32.98$$

$$\bar{y} - 109.72 + 2.6712x + 0.03935x^2$$

For example: to correct the yield of treatment 1 1.

$$Y = 79.0 - 109.72 + 2.6712(26.2) + 0.03935(699.8)$$

$$= 79.0 - 109.72 + 69.98 + 27.54$$

$$= 66.8$$

Determination of Treatment Differences

What we want to do now is determine wich treatments differ significantly from which other treatments. We will ignore the quadratic study and will use the linear regression coefficient only. We will use Tukey's HSD test and will apply it to both the uncorrected and corrected mean yields

Uncorrected Means The error mean square for the uncorrected dtabl (Table IV) is 375.9129

Treat.		Uncorrected Mean Yield	
4	3	204.2	a
3	3	191.2	ab
3	2	179.2	ab
4	2	179.0	ab
2	3	179.0	ab
4	1	176.6	ab
2	2	167.4	abc
3	1	166.2	abc
1	3	160.2	bc
2	1	133.8	c
1	2	126.4	c
1	1	79.0	d

Tukey's "w"

$$w_{.05} = q^{\alpha}(p, n_2)\ s\bar{x}$$

$$p = 12$$

$$n_2 = 44$$

$$s\bar{x} = \sqrt{375.9129/5} = 8.6707$$

$$q^{\alpha}_{.05} = 4.89$$

$$w_{.05} = 4.89 \times 8.6707 = 42.4$$

<u>Corrected</u> <u>Means</u> Corrected for linear regression only. Here we use the effective residual mean square, 135.8710

Treat.		Original Mean Yield
4	3	198.7 a
4	1	193.4 a
4	2	191.9 ab
3	3	183.8 abc
3	2	182.4 abc
2	3	182.2 abc
2	2	167.7 bcd
1	3	161.5 cd
3	1	154.9 de
1	2	130.6 ef
2	1	128.3 f
1	1	66.8 g

Tukey's "w"

$$w_{.05} = q\alpha(p, n_2)\ s\bar{x}$$

$p = 12$

$n_2 = 43$

$$s\bar{x} = \sqrt{135.8710/5} = 5.2129$$

$$q\alpha_{.05} = 4.89$$

$$w_{.05} = 4.89 \times 5.2129 = 25.5$$

If we wish to obtain a more accurate test of the significance of differences between specific corrected yields we can use the equation of Le Clerg, Leonard and Clark (page 317 of their text).

$$s_{c\bar{y}} = \sqrt{\frac{2s^2}{k} + \frac{s^2(\bar{x}_1 - \bar{x}_2)}{A}}$$

where

s^2 = the error mean square of the adjusted yield = 131.0528

k = the number of observation in each mean = 5

A = the SS for error of the independent value(stand) =465.2

$\bar{x}_1$ and $\bar{x}_2$ = the means of the independent variable (stand) for the two particular treatments being compared.

To compare treatments 1 1 and 2 3

$$s_{c\bar{y}} = \sqrt{\frac{(2x131.0528)}{5} + \frac{[131.0528(26.2-29.4)]}{465.2000}} = 7.3022$$

$$t = \frac{\bar{y}_1 - \bar{y}_2}{s_{c\bar{y}}} = \frac{66.8-18.2}{7.3022} = 15.80^{**}$$

to obtain the tabulated value of "t" look in the t Table under 44 DF.

To compare treatments 4 3 and 4 1

$$s_{c\bar{y}} = \sqrt{\frac{(2x131.0528)}{5} + \frac{[131.0528(27.6-32.2)]}{465.2000}} = 7.3291$$

$$t = \frac{198.7-193.4}{7.3291} = 0.72ns$$

THE ANALYSIS OF COVARIANCE OF A FACTORIAL EXPERIMENT

The experiment outlined above was actually a factorial with two factors, nitrogen at four levels and phosphoruos at three levels, giving twelve treatments in all. We will now illustrate the application of covariance analysis to the factorial layout. The procedure is the same as before except that the treatment sum of squares is partitiones into the three components; Nitrogen, Phosphorous and the N X P interaction. The treatment totals (from Table III) and the totals for the differents levels of Nitrogen and Phosphorous are shown in Table XIII. The sum of squares for stand and for yield and the sum of cross products for Nitrogen, for Phosphorous and for the N x P interaction are shown below Table XIII.

The analysis of variance of the uncorrected yield fata is shown below in Table XIV.

TABLE XIV

Analysis of Variance for Uncorrected Yields

Source of Variation	DF	SS	MS	F
Replicates	4	854.2333	213.5583	
Nitrogen	3	37,559.6500	12,518.8833	33.30**
Phosphorous	2	20,065.3000	10,032.6500	26.69**
N X P	6	5,984.3000	997.3833	2.65*
Error	44	16,540.1667	375.9129	
Total	59	81,003.6500		

There are significant differences between levels of nitrogen and between levels of Phosphorous. The N x P interaction is significant but only at the 5% level.

The calculation of the regression coefficient and the adjusted sum of squares for yield is shown in Table XV. The analysis of variance of the adjusted yields for nitrogen, phosphorous and the N x P interaction are shown in Tables XVI, XVII and XVIII.

TABLE XIII

Totals for the Main Effects of N and P and for the N x P Interaction

	N1		N2		N3		N4		Total	
	Std	Yld	Std	Yld	Std	Yld	Std	Yld	Std	Yld
P1	131	395	138	669	132	831	161	883	562	2778
P2	148	632	144	837	147	896	157	895	596	3260
P3	145	801	147	895	136	956	138	1021	566	3673
Tot	424	1828	429	2401	415	2683	456	2799	1724	9711

Stand

CF = 49,5363.2667
$Sum(x N)^2/n = (424^2 + ... + 456^2) / 15 - CF = 62.2667$
$Sum(x P)^2/n = (562^2 + ... + 566^2) / 20 - CF = 34.5333$
$Sum(x NP)^2/n = (131^2 + ... + 138^2) /5 - CF - N - P = 91.3333$

Yield

CF = 1,571,725.3500
$Sum(y N)^2/n = (1828^2 + ... + 2799^2)/15 - CF = 37{,}559.6500$
$Sum(y P)^2/n = (2778^2 + ... + 3673^2)/20 - CF = 20{,}065.3000$
$Sum(y NP)^2/n = (395^2 + ... + 1021^2)/5 - CF - N - P = 5{,}984.3000$

Cross Products (stand x Yield).

CF = 279,029.4000
Sum (xy N)/n = [(424 x 1828) +...+ (456 x 2799)] /15 - CF = +629.9333
Sum (xy P)/n = [(562 x 2778) +...+ (556 x 3673)] /20 - CF = +126.3000
Sum (xy NP)/n = [(131 x 395) +...+ (138 x 1021)] /5 - CF - N -P = +381.7667

Table XV
Analysis of Covariance

Source of Variation	DF	Stand SS SSX	Sum XY	Yield SS SSY
Nitrogen	3	62.2667	+ 629.9333	37,559.6500
Phosphorus	2	34.5333	+ 126.3000	20,065.3000
N X P	6	91.3333	+ 381.7667	5,984.3000
Error	44	465.2000	-2,252.3333	16,540.1667
N + Error	47	527.4667	-1,622.3333	54,099.8167
P + Error	46	499.7333	-2,026.0333	36,605.4667
(NXP) + Error	50	556.5333	-1,870.5667	22,524.4667

... continued

Source of Variation	$b_{x.y}$	$b_{x.y}SXY$	Adjusted Yield SS SSY'	New DF
Nitrogen	+10.1167	6,372.8465	31,186.8035	2
Phosphorus	+ 3.6573	461.9170	19,603.3830	1
N X P	+ 4.1799	1,595.7466	4,388.5534	5
Error	- 4.8416	10,904.8971	5,635.2696	43
N + Error	- 3.0757	4,989.8106	49,110.0061	46
P + Error	- 4.0542	8,213.9443	28,391.5224	45
(NXP) + Error	- 3.3611	6,287.1616	16,237.3051	49

TABLE XVI
Analysis of Variance for Adjusted Means of Nitrogen

Source of Variation	DF	Adjusted SS Yield	MS	F
N + Error	46	49,109.9940		
Error	43	5,635.2696	131.0528	
N1 (diff)	3	43,474.8260	14,491.6087	110.58**
N2	2	31,186.8035	15,593.4018	118.98**
bN - bE	1	12,288.0226	12,288.0226	93.76**

When we compare the F value for Nitrogen in Table XIV with the F values obtained with the adjusted sum of squares (Table XVI) we see that the F values have increased greatly, both when the sum of squares are adjusted by the error regression coefficient and the Nitrogen regression coefficient.

TABLE XVII

Analysis of Variance for Adjusted Means of _Posphorus

Source of Variation	DF	Adjusted SS Yield	MS	F
P + Error	45	28,391.5224		
Error	43	5,635.2696	131.0528	
P1 (diff)	2	22,756.3545	11,378.1772	86.82**
P2	1	19,603.3830	19,603.3830	149.58**
bP - bE	1	3,152.9715	3,152.9715	24.06**

When we compare the F values from Phosphorus in Table XIV with the F values obtained with the adjusted sum of squares (Table XVII) we see that the F values have increased greatly, both when the sum of squares are adjusted by the error regression coefficient and by the Phosphorus regression coefficient.

TABLE XVIII

Analysis of Variance for Adjusted Means of the N X P Interaction

Source of Variation	DF	Adjusted SS Yield	MS	F
NP+ Error	49	16,237.3051		
Error	43	5,635.2696	131.0528	
NP1 (diff)	6	10,602.1372	1,767.0229	13.48**
NP2	5	4,388.5534	887.7107	6.70**
bNP- bE	1	6,213.5838	6,213.5838	47.41**

In the original analysis of variance (Table XIV) the N x P interaction was significant at the 5% level. After adjusting for differences in stand the interaction is now significant al the 1% level, both when the sum of squares is adjusted by the error regression coefficient and by the interaction regression coefficient.

Alternate Procedure

In many statistic texts the analysis of covariance is presented in a different fashion. The method obtains adjusted sum of squares through the Treatment + Error component only and only the one F test is conducted. This is generally valid because the adjusted sum of squares for Treatment + Error is an estimate of the sum of squares to be expected for the two combined sources if all treatment stands were equal. You will notice that the F values obtained with the alternate proceedure are identical to the oneas obtained above. The alternate methos is shown in Table XIX.

I personally prefer the first method which we have just covered because it can provide more insight into what is actually happening. I think this will become clear in the final section.

Using the N + Error line the calculations in Table XIX are as follows:

To calculate the adjusted SS yield (SSY') for the N + Error line

$$SSY' = SSY - \frac{(SumXY)^2}{SSX} = 54{,}099.8167 - \frac{(-1{,}622.3333)^2}{527.4667}$$

$$= 54{,}099.8167 - 4{,}989.8227$$

$$= 49{,}109.9940$$

To calculate the difference for testing the adjusted N means

SSY' (for N + Error) - SSY' (for Error)

$$= 49{,}109.9940 - 5{,}635.1679$$

$$= 43{,}474.8261$$

Wo go through exactly the same steps in calculating the P + Error and the (N + P) + Error components.

TABLE XIX

Analysis of Covariance (alternate Proceedure) where Treatment

Sum of Squares is Partitioned

Source of Variation	DF	SSX	Sum XY	SSY
Replicate	4	94.4000	- 67.0667	854.2333
Nitrogen	3	62.2667	+ 629.9333	37,559.6500
Phosphorus	2	34.5333	+ 136.3000	20,065.3000
N X P	6	91.3333	+ 381.7667	5,984.3000
Error	44	465.2000	-2,252.3333	16,540.1667
N + Error	47	527.4667	-1,622.3333	54,099.8167
Difference for testing adjusted N Means				
P + Error	46	499.7333	-2,026.0333	36,605.4667
Difference for testing adjusted P Means				
(N X P) + Error	50	556.5333	-1,870.5667	22,524.4667
Difference for testing adjusted N X P Means				

... continued

Source of Variation	New DF	SSY'	MS	F
Replicate				
Nitrogen				
Phosphorus				
N X P				
Error	43	5,635.2696	131.0528	
N + Error	46	49,109.9940		
Difference for testing adjusted N Means	3	43,474.8261	14,491.6087	110.58**
P + Error	45	28,391.4632		
Difference for testing adjusted P Means	2	22,756.2953	11,378.1476	86.82**
(N X P) + Error	49	16,237.2965		
Difference for testing adjusted N X P Means	6	10,602.1286	1,767.0214	13.48**

USE OF COVARIANCE TO CORRECT FOR UNCONTROLLED GRADIENTS OR TRENDS IN EXPERIMENTS.

When we conduct an experiment which has been correctly designed, which has been correctly laid out and when we have done everything just right we expect everything to go according to Hoyle. But this does not always happen. Sometimes uncontrolled or uncontrollable factors arise which interfere with the accuracy of our results. One such factor which arises fairly commonly in field, greenhouse or growth chamber experiments is the presence of gradients across the experimental area. As an example, in a field experiment it may be found that one corner of the experimental area has given, on the average, much higher yields than the opposite corner and that there is a rather regular pattern of a reduction in yield as we go from the high corner across to the low corner. Or it may happen that the two corners are high yielding with a gradual reduction in yield toward the center giving a saddle effect.

In the field such gradients are nearly always due to differences in fertility or moisture. The same pattern(s) may be found in experiments conducted in the greenhouse or growth chambers. In these situations the gradients are usually due to temperature, light, air movement or even humidity and the gradients are usually related to distance from the edges of the bench. In laboratory experiments and in field and greenhouse experiments such gradients may exist due to time.

Such gradients may not and often do not have a noticeable effect on treatment means but such gradients do have a very pronounced effect on the experimental or residual error with the result that it is difficult or impossible to show significant differences between treatments when such differences actually are present.

When such gradients do exist it would be very nice if we could remove their effect from the analysis and there are procedures for doing just this. I will show you one such procedure which is, to me, quite simple and easily understandable in a common sense way.

Now first of all it is necessary and important to determine if a gradient does exist in your experiment and in the following account I have shown how you can detect such gradients.

An area in a field which is to be used for an experiment is shown in Table I. The lower left corner is much higher yielding than average, the upper right corner lower yielding. I have exaggerated the gradient so it will be easier for you to picture.

TABLE I
Natural Field Gradient

Rows	Replicates 1	2	3	4	5	6
1	0	-2	-4	-6	-8	-10
2	+2	0	-2	-4	-6	-8
3	+4	+2	0	-2	-4	-6
4	+6	+4	+2	0	-2	-4
5	+8	+6	+4	+2	0	-2
6	+10	+8	+6	+4	+2	0
7	+12	+10	+8	+6	+4	+2
8	+14	+12	+10	+8	+6	+4
9	+16	+14	+12	+10	+8	+6
10	+18	+16	+14	+12	+10	+8

In practically all experiments and certainly in all the ones you conduct random errors will be present. Random errors are simply a measure of the degree to which a plot (experimental unit) yields higher or lower than it's theoretical true value. The random errors in this present experiment are shown in Table II. As an example in replicate 1, row 1 the true yield for that plot should be, let us say, 24 lb. but we will record a yield of 28 lb. because the plot has a random error value of +4. In replicate 1, row 2 let us say the true yield is 26 lb. but we will record a yield of 24 lb. because the random error value is -2. Random errors are, as the word implies, distributed at random across the experimental area and do not follow a pattern like that shown in Table I

TABLE II

Random Errors

Rows	Replicates 1	2	3	4	5	6
1	+4	+1	-4	+3	+2	-2
2	-2	+2	+3	+1	-5	+3
3	-2	+1	-4	+2	+3	0
4	0	-3	+1	-2	+3	-2
5	-3	+4	-3	-1	-2	+5
6	+4	-3	+2	-5	+4	-3
7	-4	-4	-2	+1	+5	+4
8	-1	+4	+5	-1	-3	-4
9	+2	-1	+1	+3	-3	-3
10	+2	-3	-3	+3	-1	+1

When all the plot yields have been obtained the data are set up according to replicates and treatments as shown in Table III.

TABLE III

Plots Yield Including the Gradient Factor

Treatments	Replicates 1	2	3	4	5	6	Total	Mean
A	25	22	33	21	23	8	132	22
B	28	41	36	25	37	31	198	33
C	34	31	27	18	22	24	156	26
D	36	46	31	28	39	24	204	34
E	44	27	25	31	23	18	168	28
F	25	27	24	34	10	18	138	23
G	42	41	20	35	26	28	192	32
H	37	40	46	35	41	41	240	40
I	23	14	11	30	20	10	108	18
J	41	24	38	22	17	32	174	29
Total	335	313	291	279	258	234	1710	28.5

The calculation of the sums of squares and the analysis of variance is show below.

CF = 48,735.00

$S(x)^2$ = 53,992.00 - CF = 5,257.00

Reps. = 49,403.60 - CF = 668.60

Treat. = 51,042.00 - CF = 2,307.00

Source of Variation	DF	SS	MS	F
Replicate	5	668.60	133.7200	
Treatment	9	2,307.00	256.3333	5.06**
Error	45	2,281.40	50.6978	
Total	59	5,257.00		CV = 24.98%

The analysis of variance indicates that there are significant difference between treatments. Later we will conduct Turkey's test to determine whicl treatments differ from which other treatments. At the moment we ar interesed in the amount of variation throughout the experiment, that is th random error. The CV is a good measure and we see that the CV = 24.98% which is quite high and for this kind of experiment probably unacceptabl high. Our next step therefore is to determine why the CV is so high. The firs thing we will do is calculate "residuals" for each plot. In this experiment w will do so by calculating plot yield minus treatment mean yield for each plot That is X_t-$\bar{X}_t$.For example for treatment A our residuals will be +3 for replicat 1, 0 for replicate 2, +11 for replicate 3, etc. The residuals for the entir experiment are shown in Table IV.

TABLE IV

Residuals for Treatment A-J

Treatments	Replicates 1	2	3	4	5	6
A	3	0	11	-1	1	-14
B	-5	8	3	-8	4	-2
C	8	5	1	-8	-4	-2
D	2	12	-3	-6	5	-10
E	16	-1	-3	3	-5	-10
F	2	4	1	11	-13	-5
G	10	9	-12	3	-6	-4
H	-3	0	6	-5	1	1
I	5	-4	-7	12	2	-8
J	12	-5	9	-7	-12	3

We see from Table IV that the residuals vary from +16 to -14, a difference or total of 30. This is a very large range, especially when we consider that the range of treatment means is from 18 to 40, a difference of only 22. We also notice that the size of the residuals is pretty well randomly distributed throughout the table except that in replicate 1 the residuals are almost all positive and the number of negative residuals increases across the table to replicate 6 where almost all the residuals are negative. This tell us that perhaps there is a pattern to the residuals up and down as well as across the area. So we must go back to our original field layout which is shown in Table V

TABLE V

Original Field Layout

	Replicates					
Row	1	2	3	4	5	6
1	F	I	G	C	J	A
2	B	A	D	B	F	I
3	H	E	I	H	E	D
4	D	J	F	D	C	E
5	A	C	E	A	G	B
6	G	H	B	J	H	F
7	I	F	C	G	B	H
8	C	D	J	E	A	G
9	J	G	A	F	I	C
10	E	B	H	I	D	J

The next step is to insert the residuals from Table IV into their actual location in the experimental area. Thus for row 1, replicate 1 the residual for treatment F (2) is inserted; for row 2, replicate 1 the residual for treatment B (-5) is inserted, etc. The entire experimental area is handled in the same way as shown in Table VI.

TABLE VI

<u>Residual Plotted by Location in the Field</u>

Row	Replicates 1	2	3	4	5	6
1	2	-4	-12	-8	-12	-14
2	-5	0	-3	-8	-13	-8
3	-3	-1	-7	-5	-5	-10
4	2	-5	1	-6	-4	-10
5	3	5	-3	-1	-6	-2
6	10	0	3	-7	1	-5
7	5	4	1	3	4	1
8	8	12	9	3	1	-4
9	12	9	11	11	2	-2
10	16	8	6	12	5	3

When we look at Table VI we see that there are relatively large positive residuals in the lower, left-hand corner, relatively large negative residuals in the upper, right-hand corner and a noticeable gradual change in residuals as we move from one corner to the other. These large residuals, due to a definite gradient across the field are undoubtedly the major reason for the large CV and correspondingly high error variance. It would be very desirable if we could eliminate the effect of this gradient from our analysis.

Now, we cannot correct each plot yield on the basis of it's own residual. This is statistically not valid and common sense should tell you that it is not valid. What we can do however is determine how each plot might be expected to behave or should behave on the basis of the neighboring plots. How we go about this will now be shown. For replicate 2, row 2 the expected residual (adjusting residual) is -4+(-3)+(-1)+(-5) = -13. For replicate 3, 4ow 3 the adjusting residual will be -3+(-5)+1+(-1) = -8. On the edges and at the corners we use such neighboring plots as do exist and correct as follows: for replicate 1, row 2 the adjusting residual will be 4/3(+2+0-3) = -1.33 = -1. For replicate 1, row 3 the adjusting residual is 4/3(-5-1+2) = -5.33 = -5. For the corner plot, replicate 1, row 1 the adjusting residual is 2(-4-5) = -18. We continue this process for the entire experiment and end up with Table VII.

TABLE VII

Adjusting Residuals Obtained From Neighboring Plots

Row	1	2	3	4	5	6
1	-18	-13	-20	-43	-47	-40
2	-1	-13	-27	-29	-33	-49
3	-5	-15	-8	-26	-32	-31
4	-7	7	-21	-9	-27	-21
5	23	-5	8	-22	-6	-28
6	11	22	-9	6	-14	0
7	29	18	19	1	6	-7
8	39	30	27	24	5	0
9	44	43	35	28	15	1
10	40	41	41	29	23	6

It will be simplify our work from here on if we make the adjusting residuals all positive so we will add a constant (50) to each value and end up with Table VIII.

TABLE VIII

Adjusting Residuals (Constant 50 Added)

	Replicates					
Row	1	2	3	4	5	6
1	32	37	30	7	3	10
2	49	37	23	21	17	1
3	45	35	42	24	18	19
4	43	57	29	41	23	29
5	73	45	58	28	44	22
6	61	72	41	56	36	50
7	79	68	69	51	56	43
8	89	80	77	74	55	50
9	94	93	85	78	65	51
10	90	91	91	79	73	56

The final adjusting residuals are shown in Table VIII, the field layout of the treatments is shown in Table V. What we will do now is set up a Table just like Table III showing the adjusting residuals rather than plots yields. The result is Table IX.

TABLE IX

Adjusting Residuals According to Treatments and Replicates

Treatments	Replicates 1	2	3	4	5	6
A	73	37	85	28	55	10
B	49	91	41	21	56	22
C	89	45	69	7	23	51
D	43	80	23	41	73	19
E	90	35	58	74	18	29
F	32	68	29	78	17	50
G	61	93	30	51	44	50
H	45	72	91	24	36	43
I	79	37	42	79	65	1
J	94	57	77	56	3	56

To this point we have a lot of Tables, 9 in all, and of these we will combine two into one Table. Our objectives are to (1) adjust or correct the treatments means on the basis of their residuals and (2) reduce the random experimental error. We will do this by means of the analysis of covariance using the adjusting residuals as the independent variable and yield as the dependent variable. We set up Table X.

TABLE X

Adjusting Residuals and Yields

Treat.	1 Res.	1 Yld.	2 Res.	2 Yld.	3 Res.	3 Yld.	4 Res.	4 Yld.
	Replicates							
A	73	25	37	22	85	33	28	21
B	49	28	91	41	41	36	21	25
C	89	34	45	31	69	27	7	18
D	43	36	80	46	23	31	41	28
E	90	44	35	27	58	25	74	31
F	32	25	68	27	29	24	78	34
G	61	42	93	41	30	20	51	35
H	45	37	72	40	91	46	24	35
I	79	23	37	14	42	11	79	30
J	94	41	57	24	77	38	56	22
Total	655	335	615	313	545	291	459	279

...continued

Treat.	5 Res.	5 Yld.	6 Res.	6 Yld.	Total Res.	Total Yld.
A	55	23	10	8	288	132
B	56	37	22	31	280	198
C	23	22	51	24	284	156
D	73	39	19	24	279	204
E	18	23	29	18	304	168
F	17	10	50	18	274	138
G	44	26	50	28	329	192
H	36	41	43	41	311	240
I	65	20	1	10	303	108
J	3	17	56	32	343	174
Total	390	258	331	234	2995	1710

Residuals = X

CF = 149,500.4167

$S(X)^2$ = 186,637.0000 - CF = 37,136.5833

Reps. = 157,661.7000 - CF = 8,161.2833

Treat. = 150,292.1667 - CF = 791.7500

Error = 28,183.5500

Treatment + Error = 28,975.3000

Yield = Y

CF = 48,735.0

$S(Y)^2$ = 53,992.0 - CF = 5,257.0

Reps. =49,403.6 - CF = 668.6

Treat.=51,042.0 - CF = 2,307.0

Error = 2,281.4

Treatment + Error = 4,588.4

Cross Products = XY

CF = 85,357.5

SXY = 94,220.0 - CF = 8,862.5

Reps. = 87,665.0 - CF = 2,307.5

Treat.= 85,629.0 - CF = 271.5

Error = 6,283.5

Treatment + Error = 6,555.0

TABLE XI

Analysis of Covariance

Source of Variation	DF	Residual SSX	SumXY	Yield SSY	byx
Replicate	5	8,161.2833	+2,307.5	668.6	+.282737
Treatment	9	791.7500	+ 271.5	2,307.0	+.342911
Error	45	28,183.5500	+6,283.5	2,281.4	+.222949
T + E	54	28,975.3000	+6,555.0	4,588.4	+.226227
Total	59	37,136.5833	+8,862.5	5,257.0	

...continued

Source of Variation	byx.SumXY	Adjusted SSY	New DF
Replicate	652.4156	16.1844	4
Treatment	93.1003	2,213.8997	8
Error	1,400.9000	880.5000	43
T + E	1,482.9180	3,105.4820	52

...continued

Source of Variation	DF	Adjusted SSY	MS	F
T + E	52	3,105.4820		
E	43	880.5000	20.4767	
Treat. (diff)	9	2,224.9820	247.2202	12.07**
Treat. (reg)	8	2,213.8997	276.7375	13.51**
$b_E - b_T$ (diff)	1	11.0823	11.0823	0.54

Increase in Precision due to Covariance.

Before the analysis of covariance was conducted the error mean square for yield was 50.6978. A comparable figure when covariance is used is the effective residual mean square.

$$\text{Effective Residual MS} = s_{y}.x^2 \left[1 + \frac{t_{xx}}{e_{xx}} \right] \quad \text{Where}$$

$s_{y}.x^2$ = new error mean square = 20.4767

t_{xx} = treatment mean square for the adjusting residuals =

= 791.7500/9 = 87.9722

e_{xx} = error sum of squares for adjusting residuals =

= 28,183.55

$$\text{Effective Residual MS} = 20.4767 \left[1 + \frac{87.9722}{28,183.55} \right] = 20.5407$$

$$\text{Increase in precision} = \frac{\text{original error MS}}{\text{effective res. MS}} = \frac{50.6978}{20.5407} = 2.47$$

Through the use of covariance the precision of the experiment has more than doubled.

<u>CORRECTING OR ADJUSTING TREATMENT YIELD MEANS.</u>

Through the use of the regression equation $Y = \bar{Y} - b(X-\bar{X})$ we can correct or adjust the original yield means where b refers to the error regression coefficient.

For example for treatment A $Y = 22 - .222949(48.00-49.92) = 22.43$

For treatment B $Y = 33 - .222949(46.67-49.92) = 33.73$

The original and adjusted mean yields for each treatment are shown in Table XII

TABLE XII
Treatment Means Adjusted
on the Basis of the Error Regression Coefficient

Treatment	Adjusting Residual Mean	Original Yield	Adjusted Yield	Difference adj.-orig.
A	48.00	22	22.43	.43
B	46.67	33	33.73	.73
C	47.33	26	26.58	.58
D	46.50	34	34.76	.76
E	50.67	28	27.83	- .17
F	45.67	23	23.98	.98
G	54.83	32	30.91	-1.09
H	51.83	40	39.57	- .43
I	50.50	18	17.87	- .13
J	57.17	29	27.39	-1.61
	49.92	28.5		

The last step will be to conduct Tukey's test on the original yield means and then on the adjusted yield means.

Original yield means. The error mean square for the original analysis is 50.6978

Treat.	Original Mean Yld.	
H	40	a
D	34	ab
B	33	ab
G	32	ab
J	29	abc
E	28	abc
C	26	bc
F	23	bc
A	22	bc
I	18	c

Tukey's "w"

$w_{.05} = q\alpha\ (p, n_2) s_{\bar{x}}$

$p = 10 \qquad n_2 = 25$

$q\alpha_{.05} = 4.74$

$s_{\bar{x}} = \sqrt{50.6978/6} = 2.9068$

$w_{.05} = 4.74 \times 2.9068 = 13.8$

Ajusted yield means. We use the effective residual mean square = 20.5407

Treat.	Adjusted Mean Yld.	
H	39.57	a
D	34.76	ab
B	33.73	ab
G	30.91	abc
E	27.83	bc
J	27.39	bc
C	26.58	bcd
F	23.98	cd
A	22.43	cd
I	17.87	d

Tukey's "w"

$p = 10$

$n_2 = 43$

$q\alpha_{.05} = 4.74$

$s_{\bar{x}} = \sqrt{20.5407/6} = 1.8502$

$w_{.05} = 4.74 \times 1.8502 = $
$= 8.77$

There are two things I must impress upon you:

1. The adjustment process we have gone through must not be influenced, in any, by the treatments. The adjustment residuals we obtained across the experimental area would have been the same no matter what the treatments were or where they were located. In other words, theoretically we would have obtained exactly the same Table VI or Table VII if no treatments had been applied and every plot had been handled in an identical manner. What I am trying to impress upon you is that if you correct or adjust data you must do it in such a way that treatment values, treatment means or treatment differences do not, in any way, influence the adjustment.

2. The second point I want to stress is that it is perfectly all right and statistically justifiable to adjust the data as we have done above. The gradient we discovered is not a recognized part of our experiment or experimental procedure. The gradient is a random error component overlaying the experiment which has drastically inflated the error mean square. You need not say "this gradient is an unfortnate occurrence but it happened and there is nothing we can do about it". As you have seen, there is something you can do about it.

There is something else I want to point out to you. The above adjustment, by means of covariance and the obtaining of adjustment residuals based on neighboring plots is best suited to experiments in which the plots are approximately square. If the plots are long and narrow, say 15 meters long and 1 meter wide like many of our potato experiments, the above technique is not so suitable. In such a case I think you should give more emphasis to the plots on either side and less to the plots at each end. As an example take the following layout, with residuals.

-3 H	-1 K	-7 F	-5	-5
+2 C	-5 B	+1 A	-6 D	-4 E
+3 J	+5	-3 G	-1	-6

15 meters

1 meter

To adjust the plot marked A we could use the following procedure:

```
(B x 2) + (D x 2) + (F x 1) + (G x 1) =
 (-10) +    (-12) +     (-7) +     (-3) = -32
```

If C were an edge plot we would correct C as follows:

```
[ (B  2) + (H  1) + (J  1) ] 6/4 =
[  (-10) +   (-3) +   (+3) ] 6/4 = (-10) 6/4 = -15
```

If H were a corner plot we would correct H as follows:

```
(K x 2) + (C x 1) x 2 =
   (-2) +    (+2) x 2 = 0
```

Another procedure would be to multiply the side plots by 3 and the end plots by 1.

A third procedure to adjust plot A could be as follows:

```
(B x 2) + (C x 1) + (D x 2) + (E x 1) + (F x 1) + (G x 1) =

 (-10) +     (+2) +   (-12) +    (-4) +    (-7) +    (-3) =

 = -34
```

Just what adjustment procedure you use will in many cases have to be worked out by you and will depend largely on the dimensions of your plots.
The situation we covered above was rather an ideal situation where there was quite a uniform gradient across the field. Sometimes the situation in the field or greenhouse is not so ideal. I want to present one of these not so ideal situations and show you how I have handled it in the potato experiments that have been conducted. Again I must impress upon you an important point. The procedure we will follow is a "semi-cooking" procedure and so you must be extremely careful not to confound treatments with the adjustments. If you do not use great care, treatment and treatment differences may be influenced by the adjustments which of course will invalidate the results you obtain from the experiment.

The original data with the analysis of variance is shown in Table XIII.

TABLE XIII

Plot Yields by Treatments and Replicates

	Replicates						
Treatment	1	2	3	4	5	Total	Mean
A	18	24	18	9	30	99	19.8
B	16	29	25	33	37	140	28.0
C	18	30	32	40	24	144	28.8
D	9	26	32	39	41	147	29.4
E	27	29	15	31	43	145	29.0
F	34	28	30	36	34	162	32.4
G	34	26	31	41	47	179	35.8
H	24	19	48	44	50	185	37.0
	180	211	231	273	306	1201	

CF = 36,060.025

$S(x)^2$ = 40,049.0 - CF = 3,988.975

$S(x\ Rep)^2\ /n$ =37,305.875 - CF = 1,245.850

$S(x\ Treat)^2\ /n$ =37,056.200 - CF = 996.175

Analysis of Variance

Source of Variation	DF	SS	MS	F	5%
Replicate	4	1,245.850	311.4625		
Treatment	7	996.175	142.3107	2.28ns	2.36
Error	28	1,746.950	62.3911		
Total	39	3,988.975		CV = 26.3%	

The F value for treatments is not significant. The CV is 26.3% which is very high. One thing we can do and will do is look at the residuals which are plotted in Table XIV.

TABLE XIV

Treatment	Replicate 1	2	3	4	5
A	-1.8	4.2	-1.8	-10.8	10.2
B	-12.0	1.0	-3.0	5.0	9.0
C	-10.8	1.2	3.2	11.2	-4.8
D	-20.4	-3.4	2.6	9.6	11.6
E	-2.0	0	-14.0	2.0	14.0
F	1.6	-4.4	-2.4	3.6	1.6
G	-1.8	-9.8	-4.8	5.2	11.2
H	-13.0	-18.0	11.0	7.0	13.0

In Table XV we have shown the field layout of the treatments in each replicate.

TABLE XV

Field Layout of Treatments

Row	Replicate 1	2	3	4	5
1	F	A	G	C	E
2	B	E	D	H	F
3	H	C	F	B	G
4	D	H	E	A	C
5	A	F	B	D	H
6	G	D	C	E	B
7	C	G	A	F	D
8	E	B	H	G	A

The plot residuals are shown according to their location in the experimental area in Table XVI.

TABLE XVI

Residuals According to Field Location

Row	Replicate 1	2	3	4	5	Total	Mean
1	1.6	4.2	- 4.8	11.2	14.0	+26.2	+ 5.24
2	-12.0	0	2.6	7.0	1.6	- .8	- .16
3	-13.0	1.2	- 2.4	5.0	11.2	+ 2.0	+ .40
4	-20.4	-18.0	-14.0	-10.8	- 4.8	-68.0	-13.60
5	- 1.8	- 4.4	- 3.0	9.6	13.0	+13.4	+ 2.68
6	- 1.8	- 3.4	3.2	2.0	9.0	+ 9.0	+ 1.80
7	-10.8	- 9.8	- 1.8	3.6	11.6	- 7.2	- 1.44
8	- 2.0	1.0	11.0	5.2	10.2	+25.4	+ 5.08

Next, as shown in Figure 3, we plot the residuals according to rows for each individual replicate. In Figure 4 we plot the residuals for each row as an average of the replicate values.

What we do now is simply examine the plot of the residuals, first the residuals as an average of the replicate values. We see that the line fluctuates with a very noticeable dip at row 4. Row 1 and row 8 are somewhat above the zero line but remember we caccot expect everything to plot perfectly. Also when we look at the plot of residuals for each replicate we see that the plots for row 1 are indeed high for replicates 1, 4 and 5, somewhat high for replicate 2 but not for replicate 3. For row 8 the residuals are high for replicates 2 and 3 but not out of the ordinary for replicates 1, 4 and 5. In other words the residuals, very roughly, can be considered to be random events. However row 4 is different. For each replicate the residual for row 4 is extremely low. Now remember what we said about gradients or trends. In the same way row 4 is a "recognizable" source of error. The very low residuals in each replicate are not due to treatments but must be due to some outside factor such as physical damage to row 4, perhaps a dead furrow, perhaps a strip compacted by equipment, etc. It would increase the precision of our experiment greatly if we could eliminate this source of error. Also the F value for treatments does not quite reach the 5% level and yet we are pretty sure that it would if we could eliminate the considerable error associated with single row. We cannot use the previous technique because there is not a gradient across the experimental area but just this one row. I feel that in a situation like this we are justified in adjusting our data in the following way:

1. The residual of row 4 as an average of all 5 replicates is 13.6 lbs. below the mean yield for the entire experiments.

2. We increase the yield of each plot in row 4 by 13.6 lbs.

3. Our corrections are shown with the original value.

Replicate	Treatment	Original Yield	Adjusted Yield
1	D	9	23
2	H	19	33
3	E	15	29
4	A	9	23
5	C	24	38

We now insert these adjusted values into Table XIII to obtain Table XVII.

TABLE XVII

Adjusted Yields According to Treatments

Treatments	1	2	3	4	5		x
A	18	24	18	23	30	113	22.6
B	16	29	25	33	37	140	28
C	18	30	32	40	38	158	31.6
D	23	26	32	39	41	161	32.2
E	27	29	29	31	43	159	31.8
F	34	28	30	36	34	162	32.4
G	34	26	31	41	47	179	35.8
H	24	33	48	44	50	199	39.8
	194	225	245	287	320	1271	

CF = 40386.025

$S(x)^2$ = 43157.00 2770.975

Rep. = 41631.865 1245.85

Treat = 41284.2 898.175

Analysis of Variance

Source of Variation	DF	SS	MS	F	5%	1%
Rep	4	1245.85	311.4625			
Treat	7	898.175	128.3107	5.32	2.39	3.42
E	26	696.95	24.1135			
T	39	2770.975				

s = 4.9105 $\quad \bar{x}$ = 31.775 $\quad$ cv = 15.4%

As we did with the gradient, we will reduce error DF by 2DF from 28 to 26.

The treatment F value is 5.32 which is now sig at the 1 % level.

Again, I emphasize that the treatments did not enter into the correction, only the row value.

I want you to notice another point which I think can be very important. In the field layout, Table XV, in no row does a treatment occur twice. I ran into this as a complicating problem with the potato fertilizer trials and so I have laid a restriction on the randomization.. That is, no treatment will appear more than once in a row. This greatly simplifies the above correction procedure.

PARTITIONING OUT DEGREES OF FREEDOM

Nothing in partitioning is mechanical; it follows that there are times when a factorial partitioning is not appropiate even through the treatments from a factorial set. Suppose for example that a soil applied herbicide is being tested for control of wild oats. There are many sets of circumstances under which the herbicide may be used by the farmers. There will be differences in soil type, soil moisture, time of application, temperature, etc. The experimenter may select four representative soil types, A, B, C, and D and apply herbicide to each. He would, of course, also have four control plots of each soil type to which the herbicide is not applied. Let us say the experiment is conducted in the greenhouse with four replicates. The results (dry matter of wild oats in gram/plot) are as follows:

TABLE I
Randomized Complete Block Experiment

Soil type	Herbicide	Replicate 1	2	3	4	Total
A	0	67	78	62	91	298
	1	45	69	56	74	244
		112	147	18	165	542
B	0	113	122	128	151	514
	1	107	119	115	146	487
		220	241	243	297	1001
C	0	93	87	84	115	379
	1	77	83	71	108	339
		170	170	155	223	718
D	0	58	54	61	83	256
	1	62	50	63	74	249
		120	104	124	157	505
Total	0	331	341	335	440	1447
	1	291	321	305	402	1319
		622	662	640	842	2766

The sums of squares are calculated as follows:

Correction Factor = (Sum x)^2/n = (2766)^2/32 = 239,086.1250

Total SS = Sum(x)^2 = ($67^2 + 78^2 + 62^2 + ... + 63^2 + 74^2$) =
263,786 - CF = 24,699.8750

Replicate SS = ($622^2 + 662^2 + 640^2 + 842^2$)/8 = 242,961.5000 - CF=
= 3,875.3750

Herbicide SS = ($1447^2 + 1319^2$)/16 = 239,598.1250 - CF = 512.0000

Soil Type SS = ($542^2 + 1001^2 + 718^2 + 505^2$)/8 = 258,289.2500 - CF= 19,203.1250

Herb. x Soil SS = ($298^2 + 244^2 + 514^2 + ... + 249^2$)/4 = 258,951.0000
- CF - Herb. - Soil Type = 149.7500

The analysis of Variance is shown in Table II

TABLE II
Analysis of Variance

Source of Variation	DF	SS	MS	F
Replicates	3	3,875.375	1,291.7917	28.27**
Herbicide	1	512.000	512.0000	11.20**
Soil Type	3	19,203.125	6,401.0417	410.08**
Herb. x Soil	3	149.750	49.9167	1.09ns
Error	21	959.625	45.6994	
Total	31	24,699.875		

There are significant differences between the two herbicide treatments and between the four soil types. The interaction is not significant. There are really is no interest attached to the interaction because the experimenter expects the herbicide to acts differently on the different soil types. What the experimentes is interested is in how the herbicide acts on each individual soil type. To answer this question he may partition the degrees of freedom and the sums of squares as follows:

	DF
Replicate	3
Soil type	3
Herbicide in soil type A	1
Herbicide in soil type B	1
Herbicide in soil type C	1
Herbicide in soil type D	1
Error	21
Total	31

The four single degrees of freedom equald the degrees of freedom for Herbicide plus Herbicide x Soil type interaction (i.e. 1+3 = 4).

This partitioning will tell the experimentes whether the herbicides is effective or not on each of the four soil types. In these circumstances, where the existence of an interaction is virtually being assumed, or not considered important, the main effect of soil types is also rather meaningless.

The partitioning of the sums of squares into individual degrees of freedom is carried out as follows:

$$\text{Herbicide in Soil Type A} = \frac{(A0 - A1)^2}{n} = \frac{(298 - 244)^2}{8} = 364.500$$

$$\text{Herbicide in Soil Type B} = \frac{(B0 - B1)^2}{n} = \frac{(514 - 487)^2}{8} = 91.125$$

$$\text{Herbicide in Soil Type C} = \frac{(C0 - C1)^2}{n} = \frac{(379 - 339)^2}{8} = 200.000$$

$$\text{Herbicide in Soil Type D} = \frac{(D0 - D1)^2}{n} = \frac{(256 - 249)^2}{8} = 6.125$$

Notice that the sum of squares for the four individual degrees of freedom is the same as the sum of the sum of squares for Herbicide plus Herbicide x Soil Type interaction.

$$364.500 + 91.125 + 200.000 + 60.125 = 661.75$$

$$512.000 + 149.750 = 661.75$$

The analysis of variance for the individual degrees of freedom is shown in Table III.

TABLE III

Analysis of Variance

Source of Variation	DF	SS and MS	F
Herbicide in Soil A	1	364.5000	7.98*
Herbicide in Soil B	1	91.1250	1.99ns
Herbicide in Soil C	1	200.0000	4.38*
Herbicide in Soil D	1	6.1250	0.13ns
Error	21	45.6964	

The results of this analysis indicate that in soil types A and C the herbicide caused a significant reduction (.05 level) in the dry matter of wild oats. In soil types B and D there was no significant effect.

PARTITIONING TREATMENT SUM OF SQUARES INTO INDIVIDUAL DEGREES OF FREEDOM.

Many experiments are planned to determine the nature of a response curve or a response surface where the levels of a factor refer to increasing amounts of the factor. The standard linear regression analysis is apllicable as far as it goes. However, when there are equal increments between successive levels of a factor a simple coding device may be used to acomplish the analysis with much less effort. In addition, simple methods have been developed for dealing with nonlinear regression through the use of coefficients of orthogonal polynomials. Orthogonal polynomials are regression equations such that each is associated with a power of the indepependent variable, for example, with x, x^2, x^3 and all are independient of one another, or orthogonal. This permits an independient computation of any contribution according to the degree of the independient variable and independent tests of the contribution.

The coefficients and divisors for orthogonal regression comparisons for up to five equally spaced treatments (levels) are given in Table I These orthogonal polynomials are not applicable to unequally spaced treatments (levels).

TABLE I
Coefficients and Divisors for Orthogonal Comparisons in Regression Equally Spaced Treatments

Number of Treatments (levels of a factor)	Degree of Polynomial	Coefficients					Divisor=$s(c_i)^2$
		T1	T2	T3	T4	T5	
2	1 linear	-1	+1				2
3	1 linear	-1	0	+1			2
	2 quadratic	+1	-2	+1			6
4	1 linear	-3	-1	+1	+3		20
	2 quadratic	+1	-1	-1	+1		4
	3 cubic	-1	+3	-3	+1		20
5	1 linear	-2	-1	0	+1	+2	10
	2 quadratic	+2	-1	-2	-1	+2	14
	3 cubic	-1	+2	0	-2	+1	10
	4 quadric	+1	-4	+6	-4	+1	70

With three levels of a factor, there are two degrees of freedom which can be partitioned into one degree of freedom associated with the linear response curve and one for the quadratic or second degree response. Since a second or higher degree curve will always go through three points, it may be more appropriate to refer to the latter as being associated with nonlinear reponse. For four levels, and additional degree of freedom is available for estimating the cubic response, which again may be better referred to as deviations from the quatratic response.

Since the individual comparisons are independent, the sum of their sums of squares equals the sum of squares for the treatment (factor) concerned, as we shall see. Each of the individual sums of squares is tested by means of the error term (error mean square). If only the linear effect is significant, we conclude that the increase in response between succesive levels of the factor is constant. The response may be poitive or negative. A significant quadratic response indicates that aparabola fits the data better, that is, accounts for significantly more of the variation between levels than a straight line; in other words, the increae or decrease for each additional increment is not constant but changes progressively. In planning experiments in which the levels of one or more factors involve equally spaced increment, it is desiderable to have the highest level above the point where the maximun response is anticipated.

The use of orthogonal polynomials will be illustrated with an experiment designed to study the effect of storage temperature and stoge time on the quality of frozen peas (a rating of 90 equals very high quality, a rating of 10 very low quality). The results of the experiment are shown in Table II. The tratment mean are shown in Table III.

TABLE II
The effect of Storage Temperature and Storage Time on Quality of Frozen Peas.

STORAGE Temp.	Time	REPLICATES 1	2	3	4	TOTAL	MEAN
0°F	2 wks	85	70	80	78	313	78.25
	4	76	76	75	77	304	76.00
	6	76	77	72	78	303	75.75
	8	77	76	75	84	312	78.00
Total		314	299	302	317	1232	77.00
10°F	2 wks	84	80	77	80	321	80.25
	4	71	71	75	75	292	73.00
	6	66	62	63	72	263	65.75
	8	54	57	54	53	218	54.50
Total		275	270	269	280	1094	68.38
20°F	2 wks	78	73	79	77	307	76.75
	4	72	66	65	80	283	70.75
	6	57	53	55	49	218	53.50
	8	19	18	28	23	88	22.00
Total		226	210	227	229	892	55.75
	2 wks	247	223	236	235	941	78.42
Wks	4	219	213	215	232	879	73.25
Total	6	199	192	190	199	780	65.00
	8	150	151	157	160	618	51.50
Repetition Totals		815	779	798	826	3218	67.04

Table III
Table of Treatments

Storage Temperature	Weeks of Storage 2	4	6	8	Total
0°F	313	304	303	312	1232
10°F	321	292	263	218	1094
20°F	307	283	214	88	892
Total	941	879	780	618	3218

$S(x)^2/n = 3218^2/48 = 10,355,524/48 = 215,740.0 = C.F.$

$S(x)^2 = 228,664.00 - CF = \quad 12923.92$

$S(x\ reps)^2/n = (815^2 + ... + 862^2)/12 = 215,845.50 - CF = \quad 105.42$

$S(x\ wks)^2/n = (941^2 + ... + 618^2)/12 = 220,703.83 - CF = 4963.7533$

$S(x\ temp)^2/n = (1232^2 + 1094^2 + 892^2)/16 = 219,395.25 - CF = 3655.17$

$S(x\ wks\ x\ temp)^2/n = (313^2 + 321^2 + ... + 88^2)/4 = 228,103.5000$
$- CF - 4963.7533 - 3655.17 = \quad 3744.4967$

$S(x\ rep.\ x\ wks)^2/n = (247^2 + 219^2 + ... + 160^2)/3 = 220,918.0000$
$- CF - 105.42 - 4963.7533 = \quad 108.7467$

$S(x\ rep\ x\ temp)^2/n = (314^2 + 275^2 + ... + 229^2)/4 = 219,530.5$
$- CF - 105.42 - 3655.17 = \quad 29.83$

The analysis of variance of the data is shown in logical steps in three tables; Table IV, V and V.

TABLE IV
Analysis of Variance of All Components

Source of Variation	DF	SS	MS	F	5%	1%
Replicates	3	105.4200	35.1400	2.00	3.16	5.09
Weeks of Storage	3	4963.7533	1654.5844	94.10**	3.16	5.09
Storage Temp	2	3655.1700	1827.5850	103.94**	3.55	6.01
Reps x Weeks	9	108.7467	12.0830	0.69	2.46	3.60
Rep x Temp	6	29.8300	4.9717	0.28	2.66	4.01
Weeks x Temp	6	3744.4967	624.0828	35.49**	2.66	4.01
Error	18	316.5033	17.5835			
Total	47	12923.9200				

The analysis of variance in Table IV indicates that there are higly significant differences between weeks storage and between storage temperatures. Also the weeks of storage x storage temperature interaction is higly significant. The replicate component and the two interactions replicates x weeks of storage and replicates x storage temperature are not significant.

Previously I had mentioned that replicate interactions are commonly non-significant and when this is so the sum of squares for the replicate interactions can be combined with the sums of squares for the error component as shown in Table V.

TABLE V

Analysis of Variance with Replicate Interaction combined with Error

Source of Variation	DF	SS	MS	F	5%	1%
Replicates	3	105.4200	35.1400	2.55	2.92	4.51
Weeks of Storage	3	4963.7533	1654.5844	119.98**	2.92	4.51
Storage Temp	2	3655.1700	1827.5850	132.53**	3.32	5.39
Weeks x Temp	6	3744.4967	624.0828	45.26**	2.42	3.47
Error	33	455.0800	13.7903			
Total	47	12923.9200				

s = 3.7135 $\bar{x}$ = 67.04 CV = 5.54%

When we do this we see that there is a noticeable reduction in the error mean square. Also the error degrees of freedom have increased from 18 to 33. Overall, it means we have a more precise estimate of the error variance and in general will be able to detect significant differences which otherwise would be non-significant.

In Table VI we have tested the two main effects, weeks of storage and storage temperature against the interaction weeks x temperature.

TABLE VI
Analysis of Variance testing Main Effects
against the Interaction

Source of Variation	DF	SS	MS	F	5%
Weeks of Storage	3	4963.7533	1654.5844	2.65	4.76
Storage Temp	2	3655.1700	1827.5850	2.93	5.14
Weeks x Temp	6	3744.4967	624.0828		

When we do this we see that the two main effects are no longer significant. In other words, this specific test, means that the effect of storage temperature on quality of frozen peas is very dependent on the lenght of time the product was stored and that the effect of time of storage depends on what storage temperature we are discussing.

The response curve for the two main effects, weeks of storage and storage temperature are shown in Figure 1. The response curves for the simple effects that is, for the effect of storage temperatures are shown in Figures 2 and 3. The response surfaces are shown in Figures 4 and 5.

We willnow partition the main effect of storage temperature into the linear and quadratic component as in Table VII. There are 2 degrees of fredom for storage temperature and we can split this into 1 degree of freedom associated with the linear component and 1 degree of freedom associated with the quadratic component.

TABLE VII
Partition of Storage Temperature Sums of Squares
by use of Orthogonal Polynomyals.

Effect	Storage Temperature and Quality Total			Sum Q	Divisor K_r	Sum of Squares	F
	0°F	10°F	20°F				
	1232	1094	892				
Linear	-1	0	+1	-340	2(16)	3612.5	261.96**
Quadratic	+1	-2	+1	- 64	6(16)	42.67	3.09ns

For the linear component:

$Q = -1(1232) + 0(1094) +1(892) = -340$

$$\frac{Q^2}{K_r} = \frac{(-340)^2}{(-1^2 + 0^2 + 1^2)16} = \frac{(-340)^2}{32} = \underline{3612.5}$$

For the quadratic component

$Q = +1(1232) -2(1094) +1(892) = -64$

$$\frac{Q^2}{K_r} = \frac{(-64)^2}{(+1^2 + (-2)^2 + 1^2)16} = \frac{(-64)^2}{96} = \underline{42.67}$$

Each Sum of Squares has 1 degree of freedom so the sum of squares is also the mean square. Values for "F" are therefore obtained by dividing each sum of squares (mean square) by the error mean square (13.7903). The analysis shows a highly significant linear effect indicating that as storage temperature increases quality decreases. The quadratic effect is not significant indicating that there is no significant departure from a straight line over the three temperatures.

It should be noticed that the sum of the sum of squares for the linear and quadratic effects (3612.5 + 42.67 = 3655.17) which is the same as the Sum of squares for temperature (3655.17).

The partitioning of the sum of squares of the <u>main effect</u> of weeks of storage into linear and q uadratic component is shown in Table VIII.

TABLE VII

Partition of Weeks of Storage Temperature Sums of Squares by use of Orthogonal Polynomyals.

Effect	Weeks of Storage and Quality Total				Sum Q	Divisor K_r	Sum of Squares	F
	2wk 941	4wk 879	6wk 780	8wk 618				
Linear	-3	-1	+1	+3	-1068	20(12)	4752.6	344.64**
Quadratic	+1	-1	-1	+1	- 100	4(12)	208.33	15.11**
Cubic	-1	+3	-3	+1	- 26	20(12)	2.82	0.20ns

For the linear component:

$$Q = -3(941) + (-1)(879) + 1(780) + 3(618) = -1068$$

The sum of squares for the linear component is:

$$\frac{Q^2}{K_r} = \frac{(-1068)^2}{[(-3)^2 + (+1)^2 + (+3)^2]\ 12} = \frac{(-1068)^2}{240} = 4752.60$$

The sum of the squares for the quadratic and cubic components are calculated in the same way.

The analysis shows a highly significant linear effect indicating that as time of storage lengthens quality decreases. This lenar component is that portion of the sum of squares attributable to the linear regression of quality on time of storage. The quadratic component measures the additional improvement due to fitting the second order polynomial. In this instance the quadratic effect is highly significant and figure 1 shows that quality decreases with increased rapidity as time of storage lengthens. The cubic component is not significant.

Again it should be noticed that the sum of the sum of squares for the linear, quadratic and cubic effects (4752.60 + 208.33 + 2.82 = 4963.75) which is the same as the sum of squares for time of storage (4963.7533).

We have now tested the linear and quadratic and cubic components of the main effects. That is we have tested the effect of storage temperature taken over all four storage periods and we have tested the effect of weeks of storage taken over all three storage temperature. Now we will test to see if there are significant linear, quadratic and cubic effects of weeks of storage at each individual storage temperature. In Table IX we show again the coefficients for the linear, quadratic and cubic effects of weeks of storage followed by the calculations and determination of the sums of squares for each storage temperature. The values are obtained from Table III (Table of Treatment Means). For example the linear effect of weeks of storage at 0^oF storage temperature equals:

$$-3(313) + (-1)(304) + 1(303 + 3(312) = -4$$

The qadratic component equals:

$$1(313) + (-1)(304) + (-1)(303) + 1(312) = +18$$

The significance of the linear, quadratic and cubic components of weeks of storage at each storage temperature is shown in Table X.

TABLE IX
Coefficients for Determining Linear, Quadratic and Cubic Effects

WEEKS OF STORAGE

	2	4	6	8
Linear effect	-3	-1	+1	+3
Quadratic effect	+1	-1	-1	+1
Cubic effect	-1	+3	-3	+1

Storage Temperature	Effect	Weeks of Storage 2	4	6	8	Linear	Quadratic	Cubic
0°F	Linear	-939	-304	+303	+936	-4		
	Quadratic	+313	-304	-303	+312		+18	
	Cubic	-313	+912	-909	+313			+2
10°F	Linear	-963	-292	+263	+654	-338		
	Quadratic	+321	-292	-263	+218		-16	
	Cubic	-321	+876	-789	+218			-16
20°F	Linear	-921	-283	+214	+264	-726		
	Quadratic	+307	-283	-214	+ 88		-102	
	Cubic	-307	+849	-642	+ 88			-12

Determination of Sums of Squares

	Linear	Quadratic	Cubic
0°F	$\frac{(-4)^2}{4(20)} = \frac{16}{80} = 0.20$	$\frac{(+18)^2}{4(4)} = \frac{324}{16} = 20.25$	$\frac{(+2)^2}{4(20)} = \frac{4}{80} = 0.05$
10°F	$\frac{(-338)^2}{4(20)} = \frac{114244}{8} = 1428.05$	$\frac{(-16)^2}{4(4)} = \frac{256}{16} = 16.00$	$\frac{(-16)^2}{4(20)} = \frac{256}{80} = 3.2$
20°F	$\frac{(-726)^2}{4(20)} = \frac{527076}{8} = 6588.45$	$\frac{(-102)^2}{4(4)} = \frac{10404}{16} = 650.25$	$\frac{(-12)^2}{4(20)} = \frac{144}{80} = 1.8$
	8016.70	686.50	5.05

The sum of the sums of squares for the linear, quadratic and cubic effects (8016.70 + 686.50 + 5.05 = 8708.25) is the same as the sum of the sums of squares for weeks for storage and the weeks x temperature interaction (4963.7533 + 3744.4967 = 8708.25)

TABLE X
Significance of the Linear, Quadratic and Cubic Effects of Weeks of Storage at each Individual Temperature

Storage Temperature	Effect	DF	Sum of Squares and Mean Square	F
0° F	Linear	1	0.20	0.01
	Quadratic	1	20.25	1.47
	Cubic	1	0.25	0.0036
10° F	Linear	1	1428.05	103.56**
	Quadratic	1	16.00	1.16
	Cubic	1	3.20	0.23
20° F	Linear	1	6588.45	477.77**
	Quadratic	1	650.25	47.15**
	Cubic	1	1.80	0.13
Error		33	13.7903	

Notice that the Degrees of freedom total up to 9, which is the same as the sum of the degrees of freedom for weeks of storage and the weeks x temperature interaction (3 + 6 = 9).

We will use the response curve in Figures 3 and the response surfaces in Figures 4 and 5 to help us interpret the results.

At 0° F storage temperature, none of the effects are significant. That is, there is no significant change in quality over the 8 week period.

At 10° F storage temperature the linear effect is highly significant indicating again that, as storage perios lengthens there is a significant loss of quality. The quadratic and cubic effects are not significant indicating that there is no significant departure from a straight line effect.

At 20° F storage temperature the linear component is highly significant indicating again that at this storage temperature as storage period lengthens there is a significant loss of quality. The quadratic component is also highly significant indicating a significant departure from a strainght line affect. At this temperature quality decreases with increaing rapidity as time of storage lenghtens.
At no temperature is the cubic effect significant.

Next we will test to see if there are significant linear and quadratic effects due to storage temperature at each individual storage period. In Table XI we show the coefficients and the calculation. For example the linear effect of storage tmperature after 2 weeks of storage equals:

$$-1(313) + 0(321) + 1(307) = -6$$

The significance of the linear and quadratic components of storage temperature at each storage period is shown in Table XII.

TABLE XI
Coefficients for Determining Linear and Quadratic Effects

	Storage Temperature		
	0° F	10° F	20° F
Linear	-1	0	+1
Quadratic	+1	-2	+1

Weeks of Storage	Effect	Storage Temperature 0° F	10° F	20° F	Linear	Quadratic
2 wks	Linear	-313	0	+307	- 6	
	Quadratic	+313	-642	+307		- 22
4 wks	Linear	-304	0	+283	- 21	
	Quadratic	+304	-584	+283		+ 3
6 wks	Linear	+303	0	+214	- 89	
	Quadratic	+303	-526	+214		- 9
8 wks	Linear	-312	0	+ 88	-224	
	Quadratic	+312	-436	+ 88		- 36

Determination of Sums of Squares

	Linear	Quadratic
2 wks	$\frac{(-6)^2}{4(2)} = \frac{36}{8} = 4.50$	$\frac{(-22)^2}{4(6)} = \frac{484}{24} = 20.1667$
4 wks	$\frac{(-21)^2}{4(2)} = \frac{441}{8} = 55.125$	$\frac{(+3)^2}{4(6)} = \frac{9}{24} = 0.3750$
6 wks	$\frac{(-89)^2}{4(2)} = \frac{7921}{8} = 990.1250$	$\frac{(-9)^2}{4(6)} = \frac{81}{24} = 3.3750$
8 wks	$\frac{(-224)^2}{4(2)} = \frac{50176}{8} = 6272.00$	$\frac{(-36)^2}{4(6)} = \frac{1296}{24} = 54.0000$
	7321.7500	77.9167

Notice that the sum of the sums of squares for thelinear and quadratic effects (7321.75 + 77.9167 = 7399.6667) is the same as the sum of the sums of the squares for storage temperature and the weeks x temperature interction (3655.17 + 3744.4967 = 7399.6667).

TABLE XII
Significance of the Linear and Quadratic Effects of Storage Temperature at each Storage Period.

Weeks of Storage	Effect	DF	Sum of Squares and Mean Square	F
2 wks	Linear	1	4.5000	0.33
	Quadratic	1	20.1667	1.46
4 wks	Linear	1	55.1667	4.00
	Quadratic	1	0.3750	0.03
6 wks	Linear	1	990.1250	71.80**
	Quadratic	1	3.3750	0.24
8 wks	Linear	1	6272.0000	454.82**
	Quadratic	1	54.0000	3.92
Error		33	13.7903	

Notice that the degrees of freedom total up to 8, which is the same as the sum of the degrees of freedom for storage temperature and the weeks x temperature interaction (2 + 6 = 8).

We will use the response curve in Figure 2 and the reponse surfaces to help us interpret the results.

After 2 weeks and 4 weeks of storage the linear and quadratic effects are not significant. That is, no significant loss of quality due to different temperatures over this period of 4 weeks.

After 6 weeks storage the linear effect is highly significant, indicating that as storage temperature increases there is a significant loss of quality. The quadratic component is not significant, indicating that there is no real departure from a straight line.

After 8 weeks storage the picture is the same. The linear effect is even more significant indicating that the loss of quality with increasing temperature is more pronounced.

So now we have calculated the linear effect of temperature, that is the downward slope of quality with increasing temperature for each storage period. The question now is, in what way does the linear effect of temperature change with the time of storage. To answer this question we can work out linear, quadratic and cubic components of the regression of the linear effect of temperature on time of storage. To do this we use the linear sums for each storage period from Table XI. The calculation is shown in Table XIII.

TABLE XIII
Calculation of the Regression of the Linear Effects of Temperature on Storage Periods

	Weeks 2	4	6	8				
	-6	-21	-89	-224	Comparison	Divisor	Sums of Squares and Mean Square	F
$Temp_L Wks_L$	-3	-1	+1	+3	-722	4x2x20	3258.02	236.26**
$Temp_L Wks_Q$	+1	-1	-1	+1	-120	4x2x4	450.00	32.63**
$Temp_L Wks_C$	-1	+3	-3	+1	- 14	4x2x20	1.23	0.09

The Temperature Linear x Weeks Linear component is significant, so is the Temp Linear x Weeks Quadratic component. In other words we can say that this interaction is significant. From this we conclude that the linear regression of temperature on quality decreases linearly as the storage period lengthens.When we look at Figure 3 we see that at 0° F length of storage is not important, at 10° F there is a decrease in quality as storage period lengthens and that at 20° F this downward trend is even more pronounced. The significant Temp. Linear x Weeks Quadratic effect indicates that there is a significant departure from this linearity. This is most noticeable at 20 °F where the deaparture from linearity is most marked.

We have already calculated the quadratic effect of storage temperature and fond it to be non-significant (Table VII). However we may be interested in determining if the quadratic effect does change significantly as the storage period lengthens. To answer this question we can work out of liner, quadratic and cubic components of the regression of the quadratic effect of temperature on the time of storage. To do this we use the quadratic sums for each storage perio from Table XI. The calculation is shown in Table XIV.

TABLE XIV
Calculation of the Regression of the Quadratic Effects of Temperature on Storage Periods

	Weeks							
	2	4	6	8			Sums of Squares	
	-22	+3	-9	-36	Comparison	Divisor	and Mean Square	F
$Temp_Q Wks_L$	-3	-1	+1	+3	- 54	4x6x20	6.08	0.44
$Temp_Q Wks_Q$	+1	-1	-1	+1	- 52	4x6x4	28.17	2.04
$Temp_Q Wks_C$	-1	+3	-3	+1	+ 22	4x6x20	1.01	0.07

None of the "F" values are significant. From this we conclude that there is no change in the quadratic effect of storage temperature over the four storage periods. In effect at no storage period is there a significant quadratic effect of temperature on quality.

In some books on statistics other methosd are used to determine the significance of the linear, quadratic and cubic components. One methos is shown in Table XV. Our values are again obtained from Table III.

The Temperature Linear x Weeks Linear component is significant, so is the Temp Linear x Weeks Quadratic component. In other words we can say that this interaction is significant. From this we conclude that the linear regression of temperature on quality decreases linearly as the storage period lengthens.When we look at Figure 3 we see that at 0^{0} F length of storage is not important, at 10^{0} F there is a decrease in quality as storage period lengthens and that at 20^{0} F this downward trend is even more pronounced. The significant Temp. Linear x Weeks Quadratic effect indicates that there is a significant departure from this linearity. This is most noticeable at 20 °F where the deaparture from linearity is most marked.

We have already calculated the quadratic effect of storage temperature and fond it to be non-significant (Table VII). However we may be interested in determining if the quadratic effect does change significantly as the storage period lengthens. To answer this question we can work out of liner, quadratic and cubic components of the regression of the quadratic effect of temperature on the time of storage. To do this we use the quadratic sums for each storage perio from Table XI. The calculation is shown in Table XIV.

TABLE XIV
Calculation of the Regression of the Quadratic Effects of Temperature on Storage Periods

	Weeks 2	Weeks 4	Weeks 6	Weeks 8				
	-22	+3	-9	-36	Comparison	Divisor	Sums of Squares and Mean Square	F
$Temp_Q Wks_L$	-3	-1	+1	+3	- 54	4x6x20	6.08	0.44
$Temp_Q Wks_Q$	+1	-1	-1	+1	- 52	4x6x4	28.17	2.04
$Temp_Q Wks_C$	-1	+3	-3	+1	+ 22	4x6x20	1.01	0.07

None of the "F" values are significant. From this we conclude that there is no change in the quadratic effect of storage temperature over the four storage periods. In effect at no storage period is there a significant quadratic effect of temperature on quality.

In some books on statistics other methosd are used to determine the significance of the linear, quadratic and cubic components. One methos is shown in Table XV. Our values are again obtained from Table III.

TABLE XV

Determination of Linear, Quadratic and Cubic Compone
and their Interactions.

Weeks of Storage

	Wk Lin. -3,-1,+1,+3	Wk Quad. +1,-1,-1,+1	Wk Cub. -1,+3,-3,+
Temp 0 ° F	- 4	+ 18	+ 2
Temp 10° F	-338	- 16	+ 16
Temp 20° F	-726	-102	- 12
Sum	-1068	-100	- 26
Temp Linear	-722	-120	- 14
Temp Quadratic	- 54	- 52	+ 22

Storage Temperature

	Temp Linear -1,0,+1	Temp Quadratic +1,-2,+1
Weeks 2	- 6	- 22
Weeks 4	- 21	+ 3
Weeks 6	- 89	- 9
Weeks 8	-224	- 36

Components		Sums of Squares and Mean Square	F
Wks Linear	$-1068^2/240$	4752.6000	344.63**
Wks Quadratic	$-\ 100^2/\ 48$	208.3333	15.11**
Wks Cubic	$-\ 26^2/240$	2.8166	0.20
Temp Linear	$-\ 340^2/\ 32$	3612.5000	261.96**
Temp Quadratic	$-\ 64^2/\ 96$	42.6667	3.09
Wks L x Temp L	$-\ 722^2/160$	3258.0250	236.25**
Wks L x Temp Q	$-\ 54^2/480$	6.0750	0.44
Wks Q x Temp L	$-\ 120^2/\ 32$	450.0000	32.63**
Wks Q x Temp Q	$-\ 52^2/\ 96$	28.1667	2.04
Wks C x Temp L	$-\ 14^2/160$	1.2250	0.09
Wks C x Temp Q	$+\ 22^2/480$	1.0083	0.07

Compare these figures with the ones showns in Tables VII, VIII, XIII and XIV.

1.1 INTRODUCTION TO QUALITY ENGINEERING

Quality of a manufactured product is total loss generated by that product to society from the time it is shipped, other than any losses caused by intrinsic functions". By loss Taguchi refers to the loss caused by the variability of the function and loss caused by harmful side effects. When a product moves from its target value then it will cause the loss even if the product lies or not within the specification limits. This Taguchi's approach is shown in figure 1.

Quality Engineering is designed to generate quality processes. Based on the aims of Quality Engineering, TAGUCHI developed an approach to the design of experiments with the objective of reducing the costs emanated from the experimentation; this approach is more practical than theoretical and is more interested in productivity and production costs than by the statistical rules. The concepts of these techniques are based on cost and savings relationships. There are some noise factors that affect the processes, and are those that cause a functional characteristic to deviate from an objective value; these are causing variability and loss of quality.

According to TAGUCHI, this loss of quality is a long-term loss of time and money for both the consumer and the manufacturer. Within quality control activities, quality engineering consists of activities aimed at reducing variability and losses.

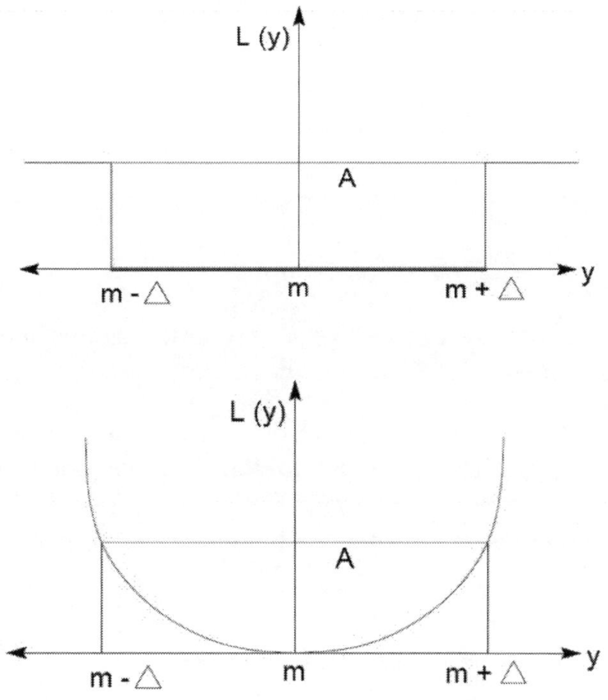

Figura 1. Traditional and Taguchi approaches to the definition of quality

Taguchi's quadratic quality loss function

Taguchi developed the concept of quality loss function in 1970. According to the Taguchi, the loss due to the performance variation is proportional to the square of the deviation the performance characteristics from its nominal value. Taguchi defined financial loss or Quality loss mathematically as:

$L(y) = k(y-m)^2$ (1)

where, y = objective characteristic or attribute

L = loss associated with y

m = specification target value

k = constant called quality loss coefficient depending upon target and specification limit

Suppose $m + \Delta$ be the specification limits and loss at $m + \Delta$ is A. Therefore,

$k = A/\Delta^2$ (2)

Here, A = cost of defective product or scrap or rework

Δ = Tolerance

Therefore equation (1) can be rewritten as

$L(y) = A/\Delta^2 (y-m)^2$ (3)

Taguchi's philosophy of robust design

Robustness may be defined as a very small or negligible variation in the performance of a product which provides consistent customer satisfaction.

Taguchi introduces his approach in using experimental design for designing a process or product in order to be robust to the environmental conditions, component variation and minimizing variation around a target value. His philosophy has far reaching consequences, yet it is founded on three very simple and fundamental concepts.

These concepts are:

(1) Quality should be designed into the product and not inspected into it.

(2) Quality should be best achieved by minimizing the deviation from a target. The product should be so designed that it is immune the uncontrollable environmental factors.

(3) The cost of quality should be measured as a function of deviation from the standard and the losses should be measured system-wide.

In first concept, Taguchi believed that better way to improve quality was to design and built it into the product. Quality improvement starts at very beginning from the design stages of a product or process which continues through the production phase. He proposed an "off-line" strategy for developing the quality improvement in place of an attempt to inspect the quality into a product on the production line. He observed that the poor quality cannot be improved by process of inspection, screening and salvaging. He emphasizes that quality is what one designs into a product.

His second concept deals with the actual method of affecting the quality. He contended that quality is directly related to the deviation of design parameter from the target value, not to conformance to some fixed specifications.

His third concept calls for measuring deviation from a given design parameter in terms of the overall life cycle cost of the products. These costs would include the cost of scrap, rework, inspection, warranty service calls and product replacement. These costs provide the guidance regarding the major parameters to be controlled.

Taguchi's three stage process to the robust design

The main objectives of the Taguchi's methods are to minimize the variation in a product response while keeping the mean response on the target and to made products robust to the changes in operating and environmental conditions.

To achieve the desirable product quality by design, Dr. Taguchi recommends a three stage process.

(1) Systems design

(2) Parameter design

(3) Tolerance design

System design is the process of applying the scientific and engineering knowledge to determine the proper working levels of selected design parameters. It includes the designing and testing a system based on the engineer's judgement of selected materials, parts and nominal product or process parameters based on current technology. Most often it involves innovation and knowledge from the applicable fields of science and technology.

Parameter design is an investigation conducted to identify the settings of design parameters that optimize the performance characteristics and reduce the sensitivity of engineering design to sources of variation or noise. So, the parameter design requires some form of experimentation for the evaluation of the effect of noise factors on the performance characteristics of the product. With the help of this experimentation optimum condition is obtained so that the influence of noise factors will be minimized and which leads to the robust design.

In this, the last step is the tolerance design which is the process of determining tolerances around the nominal setting which is already identified in the parameter design process. Tolerance design is a step used to fine tune the results of parameters design by tightening the tolerance of factors with significant influence on the product

Taguchi's methodology for single objective optimization

The following steps have been used in Taguchi Methodology (Figure 2)

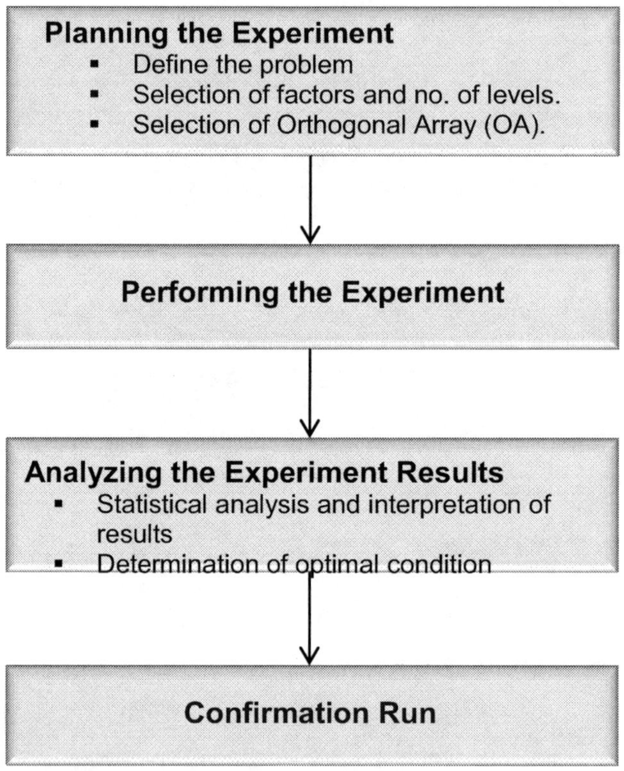

Figure 2. Steps in Taguchi Methodology for single objective optimization

A) DEFINING THE PROBLEM

It is the first step of whole process and planning phase. It involves setting the objectives of the study, defining the problem, defining the areas of concern and finally the selection of response variable or quality characteristics. The preparation required before beginning the experiment will depends on the problem under study. So careful, good and concrete planning can help in avoiding the problem that would occur in the execution of actual experimental performance. A well-defined objective ensures that experiments answers the right questions and get the better results. So, at the very first step, goals of the experiment have to be identified.

The objective is to identify those control parameters or factors settings which optimize the objective function. Developing a good and proper problem statement decides whether experiment is going in the right way or not.

At this stage it is necessary to find out the answers to these questions which are:

Why there is a need to study this problem?

Is this study would be beneficial from economical point of view or not?

Whether this study will improve the quality of the products at minimum cost or not?

After answering the above questions with the help of brainstorming and cause and effect diagram identify the objective function or quality characteristics. The objective of the experiment is to improve the performance characteristics of the product or process according to customer's needs and expectations.

The objective of experimentation should be to reduce the variation and control the variation of a product or a process and subsequently decisions must be made which parameters affect the performance of a product or process.

B) SELECTION OF PARAMETERS AND THEIR NO. OF LEVELS

In the design process, a number of factors or parameters can influence the quality characteristics or response variable of the product (output of the product) which can be classified into three classes for which the block diagram for product or process design is shown in the figure 1. These three classes are:

(1) Signal factors (M): These are the parameters set by the users or operator of the product to express the intended value for the response of the product. For example, the speed setting of a fan is the signal factor for specifying the amount of breeze; the steering angle is a signal factor that specifies the turning radius of an automobile. The signals are selected by the design engineer based on engineering knowledge of the product being developed.

Robust design projects can be classified on the basis of nature of signal factor and quality characteristics. The problems in which signal factor takes a constant value are called static problems while the other problems are called dynamic problems.

(2) Noise Factors (X): Certain parameters cannot be control by the designer and are called noise factors. The levels of the noise factors change from one unit to another, from one environment to another and from time to time. The noise factors cause the response y to deviate from the target specified by the signal factor M and lead to quality loss. Three types of noise factors can be classified as:

(i) Unit to unit noise factors: These are the random variations in a process. Examples are variation in manufacturing process, raw materials, machines, participation of operator etc.

(ii) Internal noise factors: These are the internal sources of variation in a product or process. Examples are wear and tear of components, raw material spoilage, aging, improper settings etc.

(iii) External noise factors: These are the external sources of variation in a product or process. Examples are external environmental factors like temperature, humidity, dust, load, human errors etc.

(3) Control Factors (Z): These are the parameters that can be specified freely by the designer. In fact, it is the engineer responsibility to determine the best values of these parameters. Each control factor takes multiple values which are called levels. When the levels of certain control factors are changed, the manufacturing cost also changes.

Thus, after defining the problem and selection of objective function, the next step is to identify the noise factors, testing condition and quality characteristics. The values of the various dimensions and design parameters need to be set during the initial development stages of product in order to improve the performance characteristics. List all possible parameters contributing to the problem with help of brainstorming or cause-effect diagram and the levels which each would be tested. Layout of cause and effect diagram is shown in figure 3.

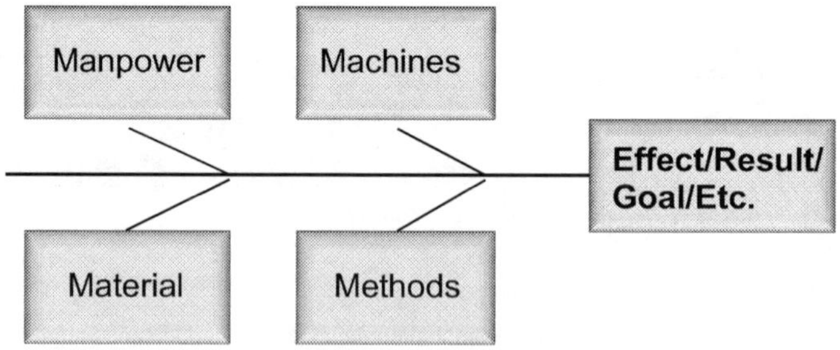

Figure 3. Layout of Cause and Effect Diagram

C) SELECTION AND USE OF APPROPRIATE ORTHOGONAL ARRAY (OA) AS THE EXPERIMENTAL MATRIX

Any process is a combination of one or more parameters and each parameter can have any value. This process will give best possible result when all these parameters operate at the optimum values. In **OA**, all parameters are varied at the same time and their effect on performance interactions are studied simultaneously. **OAs** has a balanced property that every factor levels occurs the same number of times for every level of all other factors in the experiment. A Taguchi **OA** is denoted by **LN (SM).** Here "L" denotes for Latin square, **"N"** denotes the number of experiments/trials, **"S"** denotes the levels at which these experiments is to be conducted and **"M"** denotes the number of variables/factors/parameters chosen or the number of columns in the matrix.

A "rule of thumb" is to make sure that the number of degrees of freedom associated with experiment is always greater than or equal to the number of degrees of freedom required for studying the main and interaction effects. Here, degrees of freedom are the number of fair and independent comparisons that can be made from a number of levels.

The first step of selecting the appropriate Orthogonal Array **(OA)** is to count the total degrees of freedom **(dof)**. This count fixes the minimum number of experiments that must be perform for studying the effect of parameters involved. In calculating the total dof, one dof will be given to the overall mean of the response under study; and degree of freedom for number of parameters running at different levels = Number of control parameters × (Number of Levels-1).

Therefore, the total dof = dof of overall mean of the response + dof for number of parameters or factors running at different levels. That means, one must conduct at least experiments equal to total number of **dof** to obtain the desired results. Therefore, for selecting an appropriate **OA** the number of parameters and their levels to which these parameters should be run must be known. The example of **L8(26)** orthogonal array is shown in the Table 1 and Table 2 shows the rules for selecting the standard orthogonal array (**OA**).

Table 1. Example of L8 (2^6) Orthogonal Array

Treatment Condition	Factors							Response
	A	B	C	AB	AC	BC	ABC	
1	1	1	1	2	2	2	1	*
2	2	1	1	1	1	2	2	*
3	1	2	1	1	2	1	2	*
4	2	2	1	2	1	1	1	*
5	1	1	2	2	1	1	2	*
6	2	1	2	1	2	1	1	*
7	1	2	2	1	1	2	1	*
8	2	2	2	2	2	2	2	*

Table 2. Table for the Selection of Orthogonal Array (OA)

Nu. of Levels									
	2	3	4	5	6	7	8	9	10
2	L4	L4	L8	L8	L8	L8	L12	L12	L12
3	L9	L9	L9	L18	L18	L18	L18	L27	L27
4	L16	L16	L16	L16	L32	L32	L32	L32	L32
5	L25	L25	L25	L25	L25	L50	L50	L50	L50

D) PERFORMING THE EXPERIMENTAL RUN

This is also known as implementing or conducting stage of the study. Conduct the matrix experiments as per the Taguchi's orthogonal array design matrix and then recording the responses or performance characteristics from each trial. These trials have to be closely monitored to find any discrepancies while running the experiment.

The experiment requires a minimum of one run per condition of the experiment. But one run does not represent the range of possible variability in the results. Repetition of runs enhances the available information in the data. Taguchi suggests some guidelines for the repetitions.

Repetition offers several advantages which are

(i) The additional trial data confirms the original data points.

(ii) If the noise factors vary during the day then repeating trials through the day may reveal their influence.

(iii) Additional data can be analyzed for variance around a target value.

When the cost of repetitive trials is low, repetition is highly desirable. When the cost is high or interference with the operation is high then the number of repetition should be determined by means of an expected payoff for the added cost.

Repetition permits the determination of a variation index called the signal to noise ratios (S/N). The greater this value, the smaller the product variance around the target value. The signal to noise ratios concept has been used in the fields of acoustics, electrical and mechanical vibrations, and other engineering disciplines for many years.

E) STATISTICAL ANALYSIS AND S/N RATIO CALCULATIONS

Finding a correct objective function to maximize in an engineering design problem is very important. Failure to do so can lead to great inefficiencies in experimentation and also lead to the wrong conclusions about the optimum levels. The task of finding what adjustments are meaningful in a particular problem and determining the correct S/N ratio is not always easy.

In this, the first step is to analyze the signal-to-noise (S/N) ratio, which measures the functional robustness of product or process performance in the presence of undesirable external disturbances. The S/N ratio is a special kind of data summary which is able to combine two characteristics into the desired one and is often used in analyzing the data for parameter design.

Taguchi suggests the transformation of the repetition output data in a trial into consolidated single value which is known as Signal to Noise Ratios (S/N). It may also be defined as the ratio of desirable value (signal) to the undesirable value (Noise) and it expresses the scatter around the desired value. The larger the value, the smaller will be the scatter.

The following S/N ratios are used for optimization of design parameters setting:

(1) Nominal-the-best type (NB):- In this type of problems, the quality characteristics are continuous and non-negative. For example, dimension of a part consistently achieved with modest variance.

Its target value is non zero and finite. For these problems when the mean becomes zero, the variance also becomes zero. Also, for these problems we can find a scaling factor that can serve as a adjustment factor to move the mean on the target.

The S/N ratio for nominal-the-best response is given by the following equation:

$$S/N=10\log\left(\frac{\overline{Y^2}}{S^2}\right) \quad \text{................ (1)}$$

where

$$\overline{Y}=\sum_{i=0}^{n}\frac{Y_i}{n} \qquad\qquad S^2=\sum_{i=1}^{n}\frac{(Yi-\overline{Y})^2}{(n-1)}$$

In some situations, the scaling factor can be identified by engineer expertise while in other conditions it can be identified through experimentation.

(2) **Larger-the-best type (LB)**:- In this type of problems, the quality characteristics has to be as large as possible and also it is continuous and non-negative. Also, there is no need of any adjustment factor. For example, strength of a product to be maximized, life of a component to be maximized, material removal rate during machining process etc. Since, there is no adjustment or scaling factor in these type of problems, so quality characteristics should be maximized without any adjustment.

The S/N ratios for larger-the-best response is given by the following equation

$$S/N= -10\log\left[\sum_{i=1}^{n}\frac{\frac{1}{Y_i^2}}{n}\right] \quad \text{................. (2)}$$

(3) **Smaller-the-best (SB)**:- In this type of problems, the quality characteristics has to be as small as possible and also it is continuous and non-negative. Also, there is no need of any scaling or any other adjustment factor. For example, the surface roughness during a machining process, number of defects in a product etc. Since, there is no adjustment or scaling factor in these type of problems, so quality loss should be minimized without any adjustment.

The S/N ratios for smaller-the-best response is given by the following equation

$$S/N = -10\log\left[\sum_{i=0}^{n}\frac{Y_i^2}{n}\right] \qquad \text{..................(3)}$$

Here, n represents the number of observations and Y1, Y2 ,Y3, ...Yn represents the values of a performance characteristics. Using one of the above equations according to the given condition that best fits the problem, the S/N ratio corresponding to each trial condition is computed. Now, the next step is to calculate the average S/N ratio at each level of each factor. In order to determine which of the factor/ interaction effects are statistically significant, a powerful statistical technique called analysis of variance (ANOVA) is used. Using ANOVA, one is able to identify the active and inactive factor/interaction effects with statistical confidence.

ANOVA is a mathematical technique based decision tool for detecting any differences in average performance of groups of items tested. It breaks total variation into accountable sources i.e. total variation is decomposed into its appropriate components.

ANOVA is a statistical decision tool for detecting any differences in average performance of groups of items which has to be tested. It also investigates the design parameters which significantly affect the quality characteristics.

The analysis of variance computes the quantities known as degrees of freedom, sums of squares, mean squares etc. and organizes them in a tabular form. These quantities and there interrelationships are defined as shown below using the following notations:

V = Mean of squares (Variance)

S = Sum of squares

S' = Pure sum of squares

f = Degrees of freedoms

e = Error (Experimental)

F = Variance Ratios

P = Percent contribution

T = Total (of results)

N = Number of experiments

C.F. = Correction factor

n = Total degrees of freedom

Variance: Variance may be defined as the ratio of sum of square of each trial to the degrees of freedom of each factor. Mathematically this can be express as:

$V_A = S_A/f_A$ (variance for say factor A)

$V_e = S_e/f_e$ (variance for error)

Variance Ratio (F Ratio): It is the ratio of variance of each factor to the variance for error. Mathematically this can be express as:

$F_A = V_A/V_e$

Degree of freedom: Mathematically the rank of quadratic form is known as degree of freedom. The degree of freedom can be calculated by

f_A = Number of levels of (say) factor A – 1

And degree of freedom for error variance can be calculated by

$f_e = f_T - f_A$

F) DETERMINATION OF OPTIMAL CONDITION AND PREDICTION OF PERFORMANCE

The optimal condition is the optimal parameters settings which yield the optimum performance. From the results of ANOVA, the optimal condition is obtained which yield the optimum performance. The optimal condition is obtained by identifying the levels of significant control parameters which yield the highest S/N ratios and a parameter level corresponding to the maximum average S/N ratio is called optimal level performance for that parameter and overall it is called optimal condition. Thus, the process is optimized under these conditions.

Next step is to predict the performance under these levels. With the help of optimal conditions as determined above, the optimum performance will be predicted with the help of following equation:

$$Y_{Opt} = T + \sum_{i=1}^{k}(Ti - T)$$

Here,

Y_{Opt} = Predicted optimum performance

T = Total mean of all experimental runs

T_i = Mean of all experimental runs at optimum level for factor i

K = Number of factors

G) CONFIRMATION EXPERIMENT

The last step of the Taguchi methodology is confirmation or verification experiment. The purpose of confirmation of experiment is to verify the optimum conditions so as to reduce the variation. It is very important especially when the optimum condition is not one of the trial runs already completed. Therefore confirmation experiment is conducted by using the levels of optimal setting parameters. This experimental combination of parameters resulted in substantial reduction in variation of performance characteristics and shows that the factors or parameters and levels chosen from the experiment do provide the desired results.

Confirmation experiment is the last and important step in the Taguchi process as it is the direct proof of the methodology. If the predicted and the observed values are close to each other than the used model is adequate for describing the effect of parameters on quality characteristics and if there is a large difference in observed values and predicted values then the used model is adequate.

1.2 GENERAL FACTORIAL EXPERIMENT

The results of the ANOVA for two factors can be extended to a general case where a are the levels of the factor A, b are the levels of the factor B, c are the factors of the level C, and so on, which can be arranged in a factorial experiment, in which the number of replicas is n.

It is designed to generate quality processes. TAGUCHI developed an approach to the design of experiments with the objective of reducing the costs emanated from experimentation; this approach is more practical than theoretical and is more interested in productivity and production costs than in statistical rules. The concepts of these techniques are based on cost and savings relationships.

Designing a manufacturing system to produce a product requires technical knowledge as well as extensive experience in the area to which the product belongs.

Factorial designs are widely used in experiments in which several factors intervene to study the joint effect of these on a response. There are several special cases of general factorial design that are important because they are widely used in research work, and because they form the basis for other designs of great practical value.

In recent years there has been a growing interest in some of Professor Genechi Taguchi's ideas about experimental design and its application to quality improvement

The fractional factorial design 2 k-p is used in screening experiments to quickly and efficiently identify the subset of factors that are active, and to obtain some information about the interaction. The projection property of these designs makes it possible in many cases to examine the active factors in more detail.

The sequence combination of these designs through folding is a very effective way of obtaining extra information about the interactions, which can be identified in an initial experiment as potentially important.

1.2.1 General Factorial Design 2^k

Factorial designs are widely used in experiments in which several factors intervene to study the joint effect of these on a response. There are several special cases of general factorial design that are important because they are widely used in research work, and because they form the basis for other designs of great practical value. The most important of these special cases occurs when you have k factors, each with two levels. These levels can be quantitative as would be the case of two values of temperature, pressure or time. They can also be qualitative as would be the case of two machines, two operators, the "upper" and "lower" levels of a factor, or perhaps, the absence or presence of a factor. A complete replica of such a design requires that 2 x 2 x .... x 2 = 2k observations be collected and is known as 2k general design.

The second special case is that of k factors with three levels each, known as 3k factorial design.

Be supposed to:

a) Factors are fixed

b) Designs are completely random

c) Usual assumption of normality is satisfied

The 2k design is particularly useful in the early phases of experimental work, when there are likely to be many factors to investigate. It involves the lowest number of runs with which k factors can be studied in a complete factorial design. Because there are only two levels for each factor, it must be assumed that the response is approximately linear in the range of the chosen levels of the factors.

DESIGN 2^2

The first design of the 2^k series is one that has only two factors, A and B, each with two levels. Arbitrarily, the levels of the factor can be called "inferior" and "superior".

DESIGN 2^3

Suppose that three factors A, B, and C are being studied, each with two levels. This design is known as factorial design, 2^3 and the eight combinations of treatments can be graphically represented by a cube. There are actually three different notations that are widely used for runs in the 2^k design:

a. The first is the notation "+, -", called "geometric".

b. The second is the use of lowercase letters to identify combinations of treatments.

c. In the third, the digits 1 and 0 are used to denote the high and low levels of the factor, respectively.

1.3 GENERAL FACTORIAL DESIGN 3^K

This design is a variation of the 2^k design and it is very useful as those used when all the factors act at three levels.

In recent years, there has been a growing interest in some of Professor Genechi Taguchi's ideas about experimental design and its application to quality improvement. This is a design that consists of k factors with three levels each.

The factors and interactions are represented by capital letters. The three levels of factors can be referred to as lower, intermediate and higher levels. These levels are represented by the digits 0 (lower level), 1 (intermediate) and 2 (higher). Each combination of treatments of a 3^k design is presented by k digits, where the first one affects the level of A, the second points to the level of B, ... and the kth digit, the level of the factor k.

For example, it is a design 3^2 the 00 represents a combination of treatments, in which both factor A and B are in the lower level, and 01 represents the combination of treatments corresponding to factor A in the lower level and B in the intermediate level.

In this, the system of notation that is preferred to use is + - by virtue of which it facilitates the geometric interpretation of the design and that is directly applicable to regression modeling, block formation and the construction of fractional factorials.

The addition of a third level allows modeling with a quadratic relationship the relationship between the answer and each factor.

DESIGN 3^2

The simplest design is the 32 that consists of two factors with three levels each.

Since there are 3^2 = 9 combinations of treatments, there are 8 degrees of freedom between them, The main effects A and B have two degrees of freedom each, and the AB interaction has four degrees of freedom. If there are n replicas there will be a total of $n3^2$ - 1 degree of freedom, corresponding to error 3^2 (n-1) degrees of freedom.

DESIGN 3^3

If it is assumed that three factors are being studied (A, B, C) and that each factor has three levels accommodated in a factorial experiment. This is a design 3^3. The 27 combinations have 26 degrees of freedom.

		FACTOR A		
		Low	Medium	High
		0	1	2
FACTOR B	Low 0	0	10	20
	Medium 1	1	11	21
	High 2	2	12	22

1.4 PROCESS DESIGN

Designing a manufacturing system to produce a product requires technical knowledge as well as extensive experience in the area to which the product belongs.

Mr. TAGUCHI defines quality in the following way.

"THE QUALITY OF A PRODUCT IS THE MINIMUM LOSS GIVEN BY THE PRODUCT TO THE SOCIETY, FROM THE TIME IT IS EMBARKED."

The manufacturer is the one who most resents the losses due to the negative reaction of the consumer of a product of poor quality.

FACTORIAL EXPERIMENTS

Many experiments require the study of the effects of 2 or more factors. In general, factorial experiments are the most efficient for this type of analysis. In a factorial experiment, all possible combinations of factor levels are measured in each complete stage or replica of the experiment.

When factors are arranged in a factorial experiment, they are often said to be crossed. The effect of a factor is defined as the change in the response produced by a change in the level of the factor. This is often called a main effect because it refers to the primary factors of interest in the experiment.

For example, consider the data that is represented in the following table:

Factor A

The main effect of the factor of A could be calculated as the difference between the average response of the first level of A and the average of the response in the second level of A; this is:

A = 40 + 52/2 - 20 + 30/2 = 21

This value is interpreted as the increase of factor A from level 1 to level 2 causes on average a response increase of 21 units. In the same way, the main effect of B. It is calculated below:

B = 30 + 52/2 - 20 + 40/2 = 11

In some experiments, it is found that the difference in response between the levels of one factor is not the same at all levels of the other factors; when this happens it is because there is an interaction between the factors.

For example, using the data in the following table, calculate the effect of A on the 1st level of factor B, which is done as follows:

A = 50-20 = 30

And the effect of A on the second level of factor B is:

A = 12-40 = -28

Since the effect of A depends on the selected level of factor B, we see that there is an interaction between A and B

Interaction in a Factorial Experiment

		FACTOR B	
		B1	**B2**
FACTOR A	**A1**	**20**	**40**
	A2	**50**	**12**

Graphically, both the interaction and its absence can be represented; using the data of the previous tables, two graphs will be constructed to analyze the concepts analyzed.

1.5 DESIGN OF FRACTIONAL FACTORIAL EXPERIMENTS

The number of factors in a 2^k or 3^k factorial design increased the number of runs required to perform all possible combinations of factor levels, which quickly consumed the resources available to the researchers for the experiments.

A complete replica of a design 2^6 requires 64 runs to analyze all possible combinations of treatments; in this design only 6 of the 63 degrees of freedom correspond to the main effects, only 15 degrees of freedom correspond to the interactions of two factors. The remaining 42 degrees of freedom are associated with interactions of 3 factors or more. In the 3^k series the situation is worse, for example, factorial 3^6 requires 243 runs, and only 12 of the 242 degrees of freedom correspond to the main effects.

These fractional factor designs are widely used in industrial research. An important use of fractional factorial is in research experiments, which are usually performed in the initial stages of a project, and most of the factors initially considered frequently have little or no effect on the response of the variable analyzed. The factors that are identified as important are then investigated more thoroughly in subsequent experiments.

1.5.1 Fractional Factorial Design 2^{k-p}

A 2^k fractional factorial design containing 2^{k-p} assays is known as the 1 / 2p fraction of the 2k factorial design, or simply 2^{k-p} fractional factorial design. These designs require that independent generators be chosen.

The relationship that defines these designs consists of the p generators originally chosen and their 2P - P - 1 generalized interactions. The structure of the aliases can be determined by multiplying each effect, module 2, by the relationship that defines the design. Care should be taken when choosing generators so that the effects of potential interest are not aliased with each other. Each effect has 2P-1 alias.

Typically, higher-order interactions (those of third, fourth, or greater order) are assumed negligible when the values of k are moderately large. This greatly simplifies the structure of the aliases.

1.6 Fractional Factorial Design 3k-p

Often a larger fractionation of the 3k design is desired for moderate to large values of k. In general, a fraction (1/3) P of a 3k design can be constructed for p <k, where the fraction 3k-2 is the fraction (1/9), the 3k-3 consists of selecting p interaction components, and use these effects to break down the 3k combinations of 3p block treatments. Each block now constitutes a 3k-p fractional factorial design.

The relation 1 defines any fraction, consists of the p effects initially chosen and their (3p-2p-1) / 2 generalized interactions. The aliases of any main effect or interaction component are determined by multiplying the effect by 1 e 12 modulo 3.

Example:

The maximum voltage output of a particular battery type is thought that it may be influenced by the material used in the plates and by the temperature in the location in which the battery is placed. four replicas are made in the experiment in a factorial experiment, for three types of temperature and three materials. The results are:

FACTOR (B)
TEMPERATURE

FACTOR (A) MATERIAL

1	**130** **74**	**155** **180**	**30** **80**	**40** **75**	**20** **82**	**70** **58**	**998**
2	**150** **159**	**188** **126**	**136** **106**	**122** **115**	**25** **58**	**70** **45**	**1300**
3	**138** **168**	**110** **160**	**174** **150**	**120** **139**	**96** **82**	**104** **60**	**1501**
	1738		**1291**		**770**		**3799**

First the sum squares of all the samples is evaluated, less the sum of the total of rows and columns, divided by the multiplication of rows of columns * # of samples

n = 4 A = 3 B = 3

$$SSA = \sum_{i=1}^{a} \frac{T_i^2}{bn} - \frac{\Sigma T^2}{abn}$$

$$SSA = \frac{(998^2 + 1300^2 + 1501^2)}{3*4} - \frac{(3799)^2}{3*3*4} = 10683.72$$

Now add the totals to the square of the factor, B:

$$SSB = \sum_{j=1}^{b} \frac{T_j^2}{an} - \frac{\Sigma T^2}{abn}$$

$$SSB = \frac{(1738^2 + 1291^2 + 770^2)}{3*4} - \frac{(3799)^2}{3*3*4} = 39118.72$$

Now we add the four samples of each combination and evaluate them equal. That is the sum of squares of the intersection AB example: 130 + 155 + 74 + 180 = 539

$$SSAB = SS_{ST} - SS_A - SS_B + SS_T$$

$$\frac{(539^2 + 229^2 + \dots 342^2)}{4} - \frac{(998^2 + \dots 1501^2)}{3*4} - \frac{(1738^2 + \dots 770^2)}{3*4} + \frac{(130^2 + \dots 60^2)}{3*3*4} =$$

SSAB=9613.77

Last sums of squares for the error

SSE= SST-SSA-SSB-SS(AB)

SSE= 77646.90-10683.72-39118.72 -9613.77 =18230.74

The mean square for each variation cause can be calculated as follows:

MSA =SSA/ (a-1) MSA= 10683.72/(3-1) =5341.86

MSB= SSB/ (b-1) MSB= 39118.72/ (3-1)= 19558.36

MSAB = SSAB/(a-1)(b-1) MSAB= 9613.77/(4) = 2403.44

MSE =SSE/ ab(n-1) MSE= 18230.74/ 3*3*3 =675.21

F value calculation

Fo=MSA/MSE	Fo=MSB/MSE	Fo=MSAB/MSE
Fo=5341.86	Fo=28.97	Fo=3.56

Test the Hypothesis for each factor and interaction, as well

A = Ho = 0
H1 ≠ 0

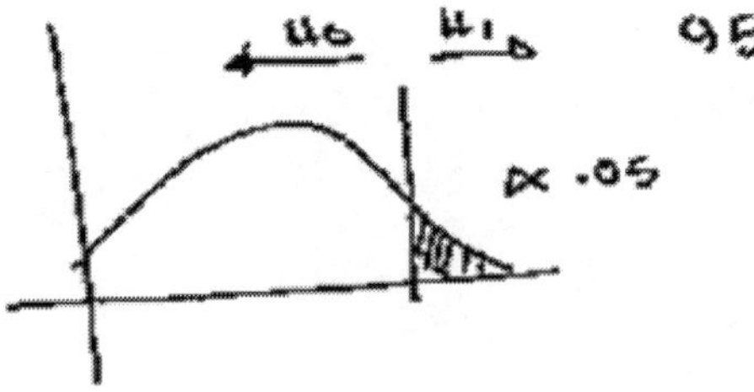

F▲ 4 9 = (.05 , 2, 27) = 3.35
F▲B = (.05 4 27) = 2.73

ANOVA TABLE FOR TWO FIXED FACTORS				
SOURCES OF VARIATION	SUM SQUARES	DEGREES OF FREEDOM	MEAN SQUARE	F vALUE
A	SSA	a - 1	MSA	MSA/MSE
B	SSB	b - 1	MSB	MSB/MSE
INTERACTION	SSAB	(a-1) 8b-1)	MSAB	MSAB/MSE
ERROR	SSE	ab (n-1)	MSE	
TOTAL	SST	abn-1		

The ANOVA table

Cause of Variation	Sums Squares, SS	Degrees of freedom, df	Mean Square, MS	F value
Material type	10683.72	2	5341.86	7.91 > 3.35
TEMPERATURE	39118.72	2	19559.36	28.97 > 3.35
INTERACTION	9613.77	4	2403.44	3.56 > 2.73
ERROR	18230.75	27	675.21	
TOTAL	77646.96	35		

CONCLUSIONS:

a. The data reflects sufficient evidence that the type of material is affected at the output voltage, considered at the level of significance of 5%

b. The data present sufficient evidence that the temperatures used in the experiments if they affect the output voltage, considering a level of significance of 5%

c. The data show sufficient evidence on the type of material used and the temperatures considered in the experiment have a joint effect on the output voltage considered ***k**, accepting the alternative Hypothesis (H1) for each case at* a significance level of 5%.

2.1 ORTOGONAL ARRANGEMENTS

This experimentation seeks to find what is the best material, the best pressure, the best temperature, the best chemical formulation, or the most suitable cycle time, etc. All for the purpose of achieving the length, breadth, or durability that is desired, taking the cost involved.

What is the orthogonal arrangement?

The orthogonal arrangement is an engineering tool that simplifies and in some cases eliminates a large part of the statistical design efforts. It is a way to simultaneously examine many factors at low cost. Dr. Taguchi recommends the use of orthogonal arrays to make matrices that contain controls and noise factors in the design of experiments. It has simplified the use of this type of design by incorporating orthogonal arrays and line graphs, finally, in contrast to traditional approaches as noise equivalents: as long as the interactions are relatively smooth, the principal effects analyst will provide us with optimal conditions and good reproducibility in an experiment. Orthogonal arrays are tools that allow the engineer to evaluate how robust the process and product designs are with regard to noise factors.

Dr. Taguchi's method for designing experiments uses techniques that involve low costs and that are applicable to the problems and requirements of modern industry. The purpose of the product design is to find that combination of factors that provide the most stable and reliable performance at the lowest manufacturing price.

The analysis of the orthogonal arrangement of Taguchi is used to produce the best parameters for the optimal design of the process, with the minimum number of experiments (tests). The results obtained for the orthogonal arrangements are analyzed to obtain the following objectives:

a) Estimate the contribution of individual factors that influence quality at the stage of product design.

b) To gain the best condition for a process or a product, so the characteristics in a good quality can be sustained.

Orthogonal arrangements and its advantage. The advantage of orthogonal arrays is that they can be applied to the experimental design involving a large number of factors.

Disadvantages. The disadvantage of the orthogonal arrangement is that it can only be applied in the initial stage of the system design of the product or process. An orthogonal arrangement allows to assure that the effect of "B" in "A1" is the same effect of "B" in "A2". This way you can be sure that you are making comparisons between the effects of one-factor levels.

ORTHOGONAL ARRAY THAT REPRESENTS $L(b)^C$ WHERE:

L = Indicates that it is an orthogonal array

a = Number of experimental runs

b = Number of levels for each factor

c = Number of columns or factors of an orthogonal arrangement.

ORTHOGONAL ARRAY $L_8(2)^7$

	1	2	3	4	5	6	7	
Nu	A	B	C	D	E	F	G	RESULTS
1	1	1	1	1	1	1	1	Y1 LEVEL 1
2	1	1	1	2	2	2	2	Y2 LEVEL 1
3	1	2	2	1	1	2	2	Y3 LEVEL 1
4	1	2	2	2	2	1	1	Y4 LEVEL 1
5	2	1	2	1	2	1	2	Y5 LEVEL 2
6	2	1	2	2	1	2	1	Y6 LEVEL 2
7	2	2	1	1	2	2	1	Y7 LEVEL 2
8	2	2	1	2	1	1	2	Y8 LEVEL 2

Orthogonal Arrays $\mathbf{2^n}$

Nu\Col	1	2	3
1	1	1	1
2	1	2	2
3	2	1	2
4	2	2	1
	Group 1	Group 2	Group 2

L_4 Graphic Array

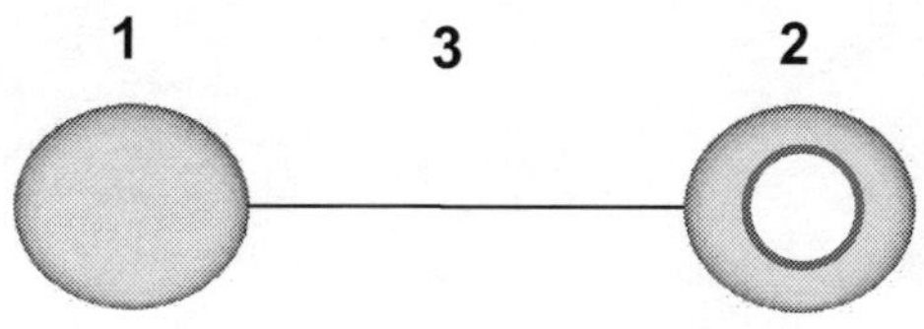

An orthogonal array is a table of combinations of the levels of orthogonally ordered results. The convention used to name orthogonal arrays is La (bc), where:

L = index that is an orthogonal array.

a = Number of experimental runs.

b = Number of levels for each factor.

c = Number of columns or factors of an orthogonal arrangement.

Orthogonal Array 2^n

Nu/Col	1	2	3	4	5	6	7
1	1	1	1	1	1	1	1
2	1	1	1	2	2	2	2
3	1	2	2	1	1	2	2
4	1	2	2	2	2	1	1
5	2	1	2	1	2	1	2
6	2	1	2	2	2	2	1
7	2	2	1	1	2	2	1
8	2	2	1	2	1	1	2

Nu/Col	1	2	3	4	5	6	7
	(1)	3	2	5	4	7	5
		(2)	1	6	7	4	5
			(3)	7	6	5	4
				(4)	1	2	3
					(5)	3	2
						(6)	1
							(7)

2^n Graphic Array

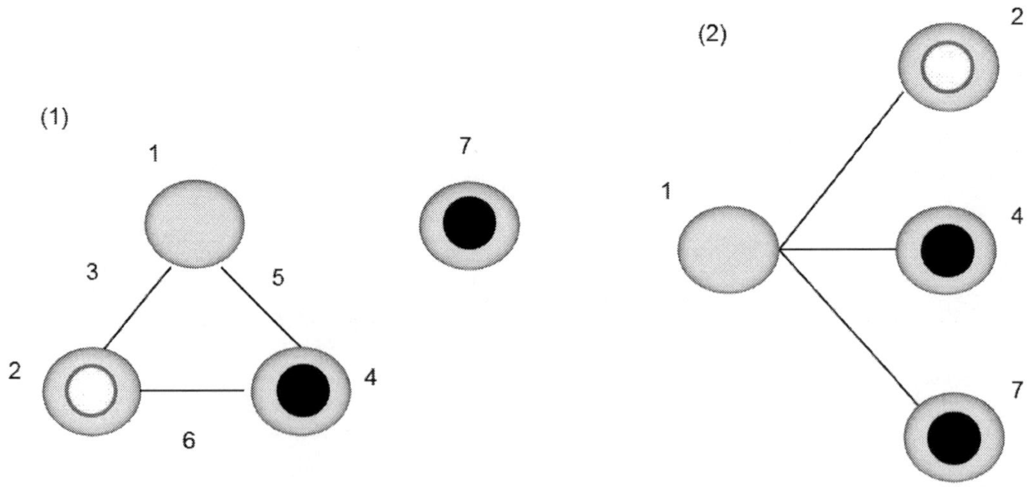

Orthogonal Array 2^n

No/Col	1	2	3	4	5	6	7	8	9	10	11	12	13	14	15
1	1	1	1	1	1	1	1	1	1	1	1	1	1	1	1
2	1	1	1	1	1	1	1	2	2	2	2	2	2	2	2
3	1	1	1	2	2	2	2	1	1	1	1	2	2	2	2
4	1	2	2	2	2	2	2	2	2	2	2	1	1	1	1
5	1	2	2	1	1	2	2	1	1	2	2	1	1	2	2
6	1	2	2	1	1	2	2	2	2	1	1	2	2	1	1
7	1	2	2	2	2	1	1	1	1	2	2	2	2	1	1
8	1	2	2	2	2	1	1	2	2	1	1	1	1	2	2
9	2	1	2	1	2	1	2	1	2	1	2	1	2	1	2
10	2	1	2	1	2	1	2	2	1	2	1	2	1	2	1
11	2	1	2	2	1	2	1	1	2	1	2	2	1	2	1
12	2	1	2	2	1	2	1	2	1	2	1	1	2	1	2
13	2	2	1	1	2	2	1	1	2	2	1	1	2	2	1
14	2	2	1	1	2	2	1	2	1	1	2	2	1	1	2
15	2	2	1	2	1	1	2	1	2	2	1	2	1	1	2
16	2	2	1	2	1	1	2	2	1	1	2	1	2	2	1
	Grupo 1	Grupo 2		Grupo 3				Grupo 4							

Two Columns Interaction

Nu/Col	1	2	3	4	5	6	7	8	9	10	11	12	13	14	15
	(1)	1	1	1	1	1	1	1	1	1	1	1	1	1	1
		(2)	1	1	1	1	1	2	2	2	2	2	2	2	2
			(3)	2	2	2	2	1	1	1	1	2	2	2	2
				(4)	2	2	2	2	2	2	2	1	1	1	1
					(5)	2	2	1	1	2	2	1	1	2	2
						(6)	2	2	2	1	1	2	2	1	1
							(7)	1	1	2	2	2	2	1	1
								(8)	2	1	1	1	1	2	2
									(9)	1	2	1	2	1	2
										(10)	1	2	1	2	1
											(11)	2	1	2	1
												(12)	2	1	2
													(13)	2	1
														(14)	1

3.0 Factores de Control y sus Límites

FACTOR	DECRIPCIÓN	NIVEL 1	NIVEL 2
A	Cantidad de Piedra Caliza	5 %(nuevo)	1 % (existen)
B	Fineza del Aditivo	Grueso (existente)	Fino (nuevo)
C	Cantidad del Aglutinante	43% (nuevo)	33% (existente)
D	Tipo del Aglutinante	Comb. (existente)	Comb.(nueva)
E	Carga de Materia Prima	1300 kg. (nueva)	1200 (existente)
F	Cantidad de Desperdicio	0% (nueva)	5% (existente)
G	Cantidad de feldespato	9% (nueva)	5% (existente)

L8 (27)

Número	A	B	C	D	E	F	G	Resultado
	1	2	3	4	5	6	7	
1	1	1	1	1	1	1	1	y1
2	1	1	1	2	2	2	2	y2
3	1	2	2	1	1	2	2	y3
4	1	2	2	2	2	1	1	y4
5	2	1	2	1	2	1	2	y5
6	2	1	2	2	1	2	1	y6
7	2	2	1	1	2	2	1	y7
8	2	2	1	2	1	1	2	y8

Each row gives a result under a different set of conditions. This allows comparisons of the different levels of the factors. An orthogonal design allows us to compare the levels of the factors under different conditions in the most efficient way.

L = RUNS = 8
2 = LEVELS
7 = FACTORS

Experiments performed in a tile factory

In A regular size tile company bought a tunnel kiln that cost $ 2 million. The oven measures 80 meters long; inside it a car loaded with tiles slowly slides on its rails while the burners bake the product.

3.1 PROBLEMS:

The tiles had a variation in their dimensions, more than 50% of the tiles in the exterior row was out of specifications, while the interior rows did comply with the measurements.

The managers and engineers knew that the causes of the variation were differences in furnace temperature. The problem could be solved by redesigning the oven so that all the tiles received the same temperature, which would have cost half a million dollars. The company did not have the resources to invest; therefore, he decided to look for ways to decrease the variation. The company did experiments to investigate the effects of some factors that in the process of cooking the tiles could affect its size. These factors were selected taking into account that the experiments were efficient.

3.2 Orthogonal arrangements series 2

An array shows the combinations of the levels of the orthogonally arranged factors. Conventionally it is represented as:

La(b)

Where:

a is the number of experimental runs.
b is the number of levels of each factor.
c is the number of columns in the orthogonal array.

The smallest orthogonal arrangement of the series (2) is L4 (2) to 3, which indicates that 3 factors are handled, with two levels each with 4 experimental runs.

Number	1	2	3
1	1	2	2
2	2	1	2
3	2	2	1

NOTE. An orthogonal arrangement that is highly recommended is the L12 (2), because it allows to investigate 11 factors without worrying about effects by interactions. This arrangement provides very good reproducibility. If an important interaction should be investigated, another major arrangement should be chosen, since it does not contemplate any interaction.

3.3 ORTHOGONAL ARRANGEMENTS OF SERIES 3

The series of three-level arrangements allows us to investigate three factors. The L9 (34) provides information of four factors at three levels, using nine experimental conditions. An L9 has eight degrees of freedom. Its use allows eight comparisons to be made orthogonally.

The eight degrees of freedom can be decomposed into two degrees of freedom per column. A column is required for each factor.

The first step is to form accumulated categories from the initial categories, so that the accumulated category 1 is equal to the initial category 1, the accumulated category 2 is equal to the initial categories 1 plus 2.

EXAMPLE:

(i) = (1)
(ii) = (1) + (2)
(iii) = (1) + (2) + (3) etc.

The second step is to know the proportion that each accumulated category has.

EXAMPLE:

P1 = 25/90
P11 = 49/90
P111 = 65/90
PIV = 90/90

Each category is assigned a weight, according to the following expression:

Wj = 1 / (Pj * (1-Pj))

For each category, the correlation factor is calculated, with the following formula:

CFj = (i¨2) / n

SUM OF FACTOR SQUARES. It is obtained by the sum of squares of each class, multiplied by its weight according to its formula.

Ssa = (Ssa class 1) * W1 + (Ssa class 11) * W11 + (Ssa class 111) * W111 .... + (Ssa class n) * Wn

Sstotal = (total number of data) * (number of categories - 1).

DEGREES OF FREEDOM. The degrees of freedom are calculated based on the degrees of freedom of a factor for variables, multiplied by the number of accumulated categories minus one.

EXAMPLE: In a problem you have, four factors with three levels we get the following

2 * (4-a) = 6

We obtain that this example has 6 degrees of freedom.

THE ERROR. The error can be obtained by subtracting the total sum of squares of each factor.

Ss error = Ss total -Ssa -Ssb -Ssc -Ssd

With all this data, an Anova Table is constructed to take a decision.

CONCLUSION. The analysis of variance offers more than a short method to obtain the sum of squares of the error. It is reliable and of great accuracy to obtain the correct option and to diminish cost by many experiments.

3.4 LINEAR GRAPHICS

They are used to facilitate the indication of factors and interactions in an orthogonal arrangement. Each point represents a column to assign some factors. The lines that connect the dots represent the columns that must be assigned to the interactions between the connecting factors.

ORTHOGONALITY. An orthogonal design allows to compare, with the same efficiency, the levels of the factors under various conditions.

DEGREES OF FREEDOM OF A FACTOR. The degrees of freedom are a measure of a quantity of information that can be obtained. To know the effect of a certain factor in an experiment, the performance of the product is compared by putting the factor in several levels when performing an experiment; the more levels a factor has, the more degrees of freedom it will have, and therefore more information to be obtained.

INTERACTION BETWEEN FACTORS. There is an interaction when the effect of a factor depends on the level in which another factor is found. The changes of a factor B will be plotted to determine if there is interaction. Graphics facilitate the allocation of factors and interactions to an ortogonal arrangement.

Its interpretation is the following:

** The numbers represent the number of the column in the orthogonal array.

** The points represent the factors.

** The line segments that join two points represent the interaction between those two factors.

3.5 INTRODUCTION OF FACTORS TO ORTHOGONAL ARRANGEMENTS

COMPLETE FACTORIAL DESIGN. In this design all combinations of all levels of all factors are investigated, allowing to investigate the effect of several factors at the same time. In an experimental design of 7 factors in 2 levels each, 128 experiments are required --- (2^7)

In manufacturing experiments it is common 13 factors with 3 levels each, being then necessary (3) 13 = 1,594,323 experiments. DATA ANALYSIS BY ORTHOGONAL ARRANGEMENTS

- ✓ Determination of response averages for factor levels.
- ✓ Selection of optimal levels of a factor by comparing averages of responses.
- ✓ Predict the average response of the process using the optimal levels.
- ✓ Comparison of the prediction with the results of a confirmation run.

INTERACTIONS BETWEEN FACTORS. There is an interaction when the effect of a factor depends on the level at which another factor is found. The changes from an "A" factor to changes in the "B" factor are gratified to see if there is interaction.

- ✓ If the lines drawn are parallel, there is no interaction between the factors.
- ✓ If the lines are not parallel, it means that the effect of "A" is not the same for "B1" and "B2", there being interaction.
- ✓ If the lines intersect in the graph, it is interpreted that there is a quite strong interaction.

APPROACHING A FACTOR AT A TIME. In this method the level of a single factor is varied, keeping the levels of the other factors constant. Suppose you investigate about temperature and pressure effects. Two levels are selected for the temperature factor (T1 and T2) and 2 levels for the pressure factor (P1, P2). The temperature is set to T1 while varying the pressure levels; then the pressure could be set to P1 and the temperature levels vary. If a design was made with 7 factors in 2 levels each, it would be necessary to perform 8 experiments.

Experiment	Factors and Conditions							Experimental Results
	A	B	C	D	E	F	G	
1	A1	B1	C1	D1	E1	F1	G1	Y1
2	A2	B1	C1	D1	E1	F1	G1	Y2
3	A2	B2	C1	D1	E1	F1	G1	Y3
4	A2	B2	C2	D1	E1	F1	G1	Y4
5	A2	B2	C2	D2	E1	F1	G2	Y5
6	A2	B2	C2	D2	E2	F1	G1	Y6
7	A2	B2	C2	D2	E2	F2	G1	Y7
8	A2	B2	C2	D2	E2	F2	G2	Y8

LIMITATIONS
· It is not practical when many factors are used
· Does not guarantee reproducibility.

3.6 MODIFICATION TO ORTHOGONAL ARRANGEMENTS

Preset based on the modifications in the line graphs.

OBJECTIVE: That these graphics can be used for a particular problem. The standard linear graphs can be modified to be used in a specific experiment, using triangular matrices, and the orthogonality between the columns is maintained.

EXAMPLE: An L16 can be modified, remaining as follows:

FACTORS: A, B, C, D, E, F, G, H, I, J.

INTERACTIONS: AxB, CxD, ExF, FxG, and GxH.

Solution:

1. The linear graph is drawn according to factors and interactions:

2. Select the standard linear graph from which the linear graph of the previous point can be obtained

3. Reviewing the triangular matrix, we determine how the linear graph can be modified if the interaction between columns 6 and 11 is needed, then column 13 is eliminated and the interaction between the columns and it is left.

SOLUTION OF THE ORTHOGONAL ARRANGEMENT.

The degrees of freedom are used to select a smaller orthogonal array that is appropriate. The required degrees of freedom are first established based on the number of factors and the number of levels of> each of these, as well as the interactions that are desired to be investigated. Example:

If you have 5 factors (A, B, C, D, E) with two levels each, and you also need to investigate the interactions AXB and AXC, it is necessary to use an orthogonal array L8, which has 7 degrees of freedom.

Example:

Degrees of freedom of the factors (5X1)	= 5
Degrees of Freedom of AXB (1X1)	= 1
Degrees of freedom of AXC (1X1)	= 1
Total	= 7

ALLOCATION OF FACTORS TO AN ORTHOGONAL ARRANGEMENT

PROCEDURES:

1. The appropriate orthogonal array is selected; for this, the degrees of freedom necessary due to the factors and interactions are taken into account.

2. Draw the linear graph according to the desired interactions.

3. One of the standard line graphs of the chosen arrangement is selected, from which the linear graph of the previous step can be obtained.

4. The factors and interactions are assigned according to the columns that correspond to them according to the resulting linear graph

EXAMPLE: Newton's Law of Motion F = MA. Nature knows this relationship but we do not. How can we get the relationship quickly without getting caught up in years of theoretical research like Newton did? A DDE provided for two inputs, such as mass (M) and acceleration (A), each at 2 levels, can be established as a process of 4 experimental trials (or attempts) in low and high positions as shown in the table.

Assay	M	A
1	LOW	LOW
2	LOW	HIGH
3	HIGH	LOW
4	HIGH	HIGH

A set of experimental conditions is used to interrogate nature about the behavior of them under certain circumstances the specific set of conditions (or design matrix) is constructed to analyze the response, with the purpose of building a model that approximates the model real finding in nature (ie, approximately equal to perfect knowledge on the part of nature). The example of the table, in DDE is known as a 2-factor complete 2-factor factorial design. Various types of experimental design are available to meet the specific objective you have, I.e. linear modeling, non-linear modeling, selection, etc. Assume that our experiment generated an average set of results for strength for each experimental trial, as shown in the table.

The fixed points, low and high, for the table (5, 10) and the acceleration (100, 200) are chosen based on ranges of interest for the experimenter

ASSAY	INPUTS		OUTPUTS (mean)
	M	A	F
1	5	100	500
2	5	200	1000
3	10	100	1000
4	10	200	2000

To maximize the amount of knowledge acquired, the modeling analysis of a DDE is conducted with standardized input values, where the low values are coded in (1) and the high values are coded in (+ 1). These coded values standardize the scale and the units of the input variables The result is a matrix with coded entries as shown in the next table.

ASSAY	M	A	M*A	F
1	-1	-1	1	500
2	-1	1	-1	1000
3	1	-1	-1	1000
4	1	1	1	2000

*Note, in the table, the addition of a column M * A which is generated by the product of the coffered columns M and A, and that was used to analyze the interactive or combined effect of M and A. The columns M, A and M * A represent the three effects to be evaluated, ie the linear effects of M and A, and the effect of the interaction of M with A (M * A). The analysis is conducted as shown in the table.*

ASSAY	M	A	M*A	F
1	-1	-1	1	500
2	-1	1	-1	1000
3	1	-1	-1	1000
4	1	1	1	2000
MEAN(+1)	1500	1500	1250	1125
MEAN(-1)	750	750	1000	
MEAN(+1)-MEAN(-1)	750	750	250	

To obtain the numbers of the shaded sections, you must do the following for each of the columns M, A and M * A: find the average of the output F when the values of the column are -1, find the average of the output F when the values of the columns are +1, and then find the difference of the two averages (Δ). For example, when the columns of the effect M is A - 1 (Tests 1 and 2) the output is 500 for the test 1 and 1000 for the test 2, which on average DA 750.

The model generated by the DDE is constructed using a minimum square regression design which can be simplified for 2-level designs as shown below:

F Is the predicted average for strength
Mc Is the variable coded for mass
Ac Is the encoded variable for acceleration
FEs the great experimental media
M Is the linear effect size for M, etc.

Then our current model becomes:

F = 1125 + (750/2) * Mc + (750/2) * Ac + (250/2) * Mc * Ac

Remember that our prediction model is for coded values

(I. E -1, +1) of M and A, and we can transform it into a model with current values of Ma and Aa using the following relationship:

Ma = ((Ma + Mb) / 2) + ((Ma + Mb) / 2) * Mc

Where:
MA Current fixed value of mass
MC Fixed value coded mass
MA Current experimental value set high
MB Current experimental value set at low

For our example:

Ma = 7.5 + 2.5 Mc

then:

The calculations previously described are a simple way to move from coded to current values and vice versa, as shown in the combined scale as shown in the next Figure.

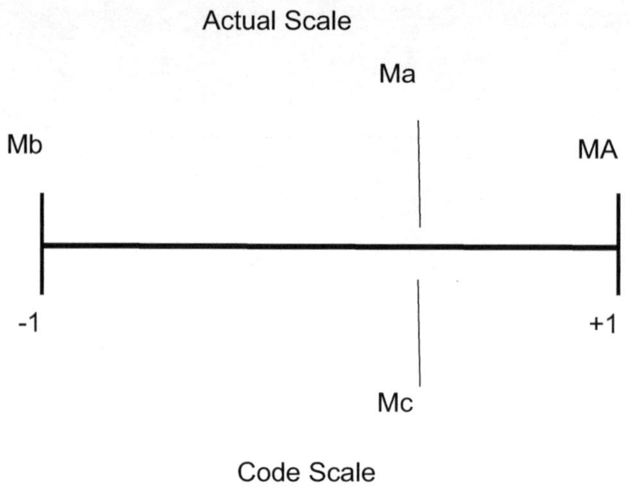

4.0 CASE STUDY

Application of Taguchi Method in Riveting Process

Jaharah A. Ghani, Haris Jamaluddin, Rizauddin Ramli, Mohd Nizam Ab Rahman, Baba Md Deros and Che Hassan Che Haron Department of Mechanical and Materials Engineering, Faculty of Engineering and Built Environment Universiti Kebangsaan Malaysia, 43600 Bangi, Malaysia

Introduction

In the recent market and global requirements, the desire to provide higher quality level of products and services to increase market shares has continued competitively among manufacturers to ensure customer satisfaction and assure consumer return orders. The importance of product and process optimization has prompted manufacturers to focus greater attention to design optimization to stay competitive in the world market. Conventional design of experiment (DOE) has been known to be a very complicated and a highly disciplined process, which requires competent resources and demands high experimental cost. It is complex and not easy to implement, especially when a large number of experiments need to be conducted when the number of process parameters increases [1]. The time required to complete an experiment is extremely long, especially in investigating and evaluating a large quantity of factors that affects the desired quality characteristics. Difficulties are further aggravated when the experiments have to be repeated for several modeling and verification purposes until accurate and validated results are obtained. Therefore, the DOE Taguchi method has become an alternative in solving these problems and is chosen as an appropriate solution for any industrial organization in improving its product and process design. The Taguchi method, designed to reduce engineering experimental time and cost, stimulates the initiative and effort for product improvement and assists in continuous improvement of processing capability.

The practicability of DOE, with its simplified data collection, has made its practical implementation possible in any organization and business operation for product and process improvement. The dilemma confronting most manufacturing industries is the lack of competent and skilled personnel to implement DOE, also a similar problem with the Taguchi method, although it has been the most recommended method for application in the industry. In most manufacturing industries, only a handful of personnel may be capable of implementing DOE.

Nevertheless, organizational policies kept on highlighting the importance of DOE in product and process design. Some organizations, which have realized the importance of product improvement to stay competitive in the market place, have capitalized on producing better quality products at lower cost and shorter period for new product development, as well as started investing in human capital by educating and improving their employees' skills and knowledge in DOE. The Taguchi method emphasizes the importance of designing quality control in the manufacturing processes [2].

Robust design is an engineering methodology employed to optimize product and process conditions, which are minimally sensitive to various causes of variations and which produce high product quality with low development and manufacturing costs [3].

The Taguchi parameters and tolerance design are important tools for robust design. Robust design and, indeed, any statistical analysis need to be done to conform to the engineering aims of the study [4]. The design process in real and feasible DOE is a combination of statistical theorem and engineering practices.

The overall target of quality engineering is to make products that are robust with respect to all noise factors [5]. According to Apte [6], the variations that represent the noise factors are categorized under the following three sources:

i) Unit to unit variation: caused by the variation in material, equipment performance, and workmanship

ii) Environment (external factor): arises from environmental conditions surroundings the processes

iii) Deterioration (internal factor): wear and tear of the components and expendable items used in the process

Often, in an industrial setting, totally eliminating the source of variations can be very expensive [2]. A low-cost solution can be achieved by adjusting the levels of variation. Figure 1 shows the Taguchi design procedure that represents the three stages: system, parameter, and tolerance designs [7]

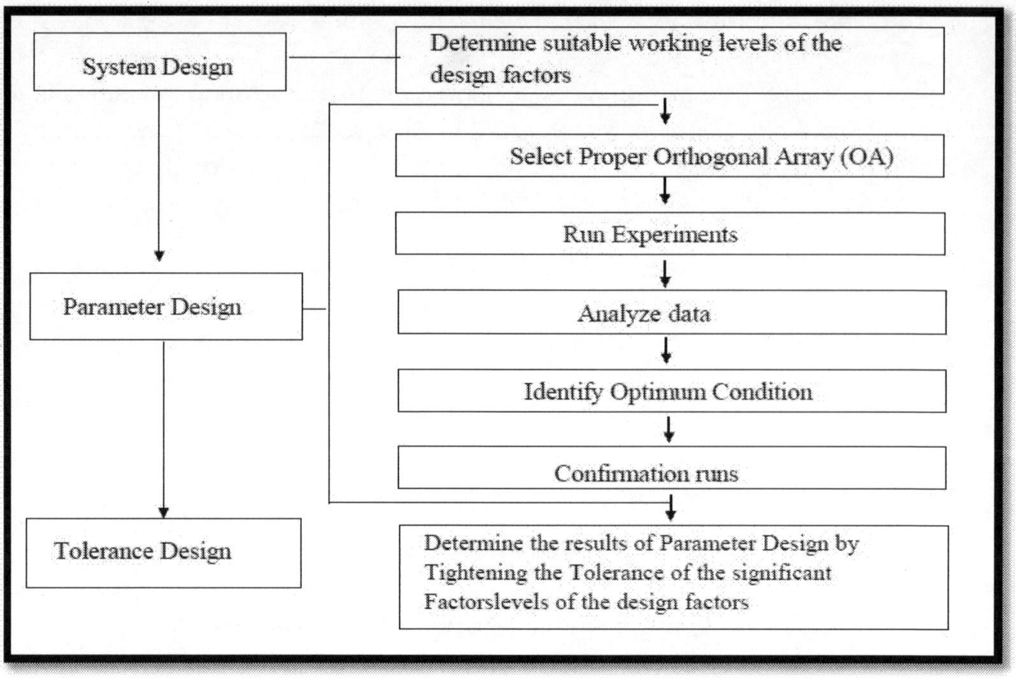

Figure 1: Taguchi design procedure

2. Methodology

The effectiveness of the Taguchi method has been verified by the procedures and methodology used in this approach, where experiments were conducted for a static case in the production line. The experimentation was based on dual-response variables for "nominal-the-best" and "larger-the-better" quality characteristics in a precision riveting process. In this case study, only the "larger-the-better" and "nominal-the-best" characteristics were studied in the experiment. The process started with the identification of the problem nature that occurred at the workplace. To execute the experiments, the related quality characteristic was identified for project selection, and the current performance of this quality characteristic was used as a baseline for comparison. The next stage was the identification of the factors that can possibly affect the quality performance of the response variable. Each factor level was tested and determined to select the appropriate Taguchi orthogonal array. In this study, the Taguchi L9 array was selected as the best array. The noise factors were identified and included in the experiment, as their potential effects on the performance characteristic could be significant. The experiment was then conducted at the actual production floor according to the sequence of the Taguchi L9 array to capture all potential variations from the controllable and uncontrollable factors during the process. The results of the experiments during the observation period were recorded in matrix columns.

2.1 Riveting Process: Dual-response Quality Characteristics

The riveting process in this case study was implemented using a radial riveting machine, which applied radial force to press the guide shaft to form a snap head that will hold the shaft to the lever plate. Figure 2 shows the radial riveting machine used to join the pin and the plate lever, and Fig. 3 shows the schematic diagram of the radial riveting process and the workpiece. The response variables are riveting strength (N·cm) and riveting snap height (mm). The expected minimum force required for this riveting strength was specified at 4 N·cm by the customer, and the riveting snap height was set from 0 mm to 0.9 mm. The guide shaft is made from SUS (SF20x) steel, and the lever is made from SUS430-Cx steel.

Figure 2: Radial riveting machine used to join the pin and plate lever

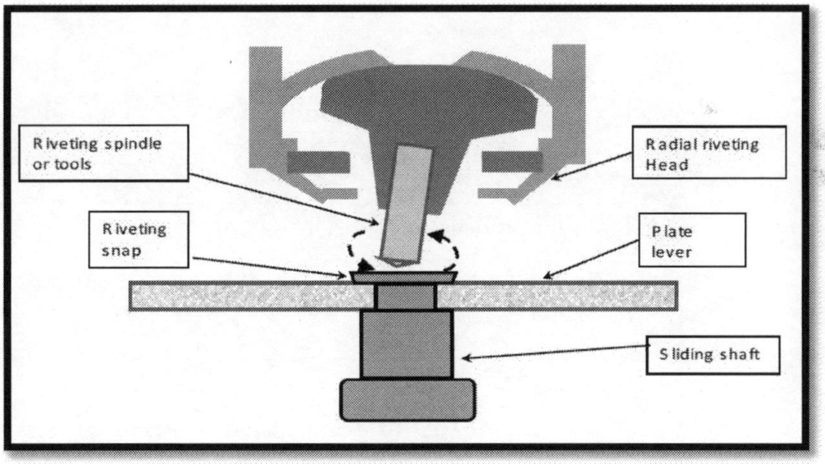

Figure 3. Schematic diagram of the radial riveting process and the work piece

Radial forming displaces a material in a uniform way, at a constant speed, and in three directions, namely, radially outward, radially inward, and tangentially overlapping. The tool moves in an overlapping radial motion pattern, causing gentle displacement of the material. The experiment was conducted employing two noise factors: thickness of the plate and height of the shaft, which were produced by external source. The thickness and height of each noise factor were classified into high and low values. The relationship among the controllable factors, the noise factors, and the quality characteristics in the riveting process is shown in Fig. 4

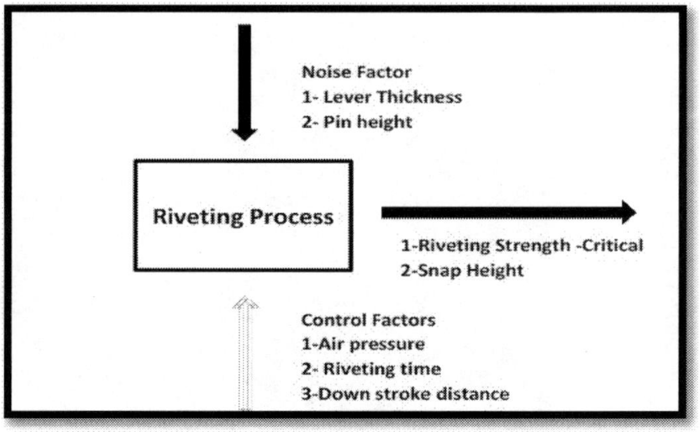

Figure 4. Block diagram of the riveting process for a static problem

The control factors and their levels tested in the experiment are shown in Table 1.

Table 1. Control factors and their levels for the riveting experiment

Factor	Units	Level 1	Level 2	Level 3
A. Cylinder air pressure	MPa	0.3	0.4	0.5
B. Caulking time	Second	1.5	2.0	2.5
C. Stroke distance	mm	2.0	2.05	2.15

The combination setting of the experiments used for the riveting strength and snap height is shown in Table 2.

Table 2. The L9 array with combination setting and result column used in riveting process

	Control Factors				Noise Factors				
Expt. No.	A	B	C	empty	T1H1	T1H2	T2H1	T2H2	S/N
1	0.3	1.5	2	1	Result1	Result2	Result3	Result4	
2	0.3	2	2.05	2	Result1	Result2	Result3	Result4	
3	0.3	2.5	2.15	3	Result1	Result2	Result3	Result4	
4	0.4	1.5	2.05	3	Result1	Result2	Result3	Result4	
5	0.4	2	2.15	1	Result1	Result2	Result3	Result4	
6	0.4	2.5	2	2	Result1	Result2	Result3	Result4	
7	0.5	1.5	2.15	2	Result1	Result2	Result3	Result4	
8	0.5	2	2	3	Result1	Result2	Result3	Result4	
9	0.5	2.5	2.05	1	Result1	Result2	Result3	Result4	

A = Air pressure
B = Riveting time
C = Stroke Distance

Noise factor i) T- Lever thickness 1 and 2.
Noise factor ii) H- Pin Height 1 and 2.

The measurement of the quality characteristics was done using a manual torque meter for the strength and a digital indicator for the snap height.

3. Results and Discussions

The riveting process for the "larger-the-better" and "nominal-the-best" characteristics was chosen for optimum parameter setting. The Taguchi method was employed in the experiments for selected quality characteristics. The experiment was conducted based on L9 array with nine trials, repeated four times for the two noise factors at two levels. In the riveting process, the L9 array was used to determine the effect of three process parameters: A—cylinder air pressure; B—riveting time; and C—down stroke. The two noise levels were identified as thickness of the lever and height of the guide shaft. The result of the experiment for the "larger-the-better" quality characteristic is shown in Table 3.

Table 3. The result of the riveting strength

	FACTORS			NOISE FACTOR						
RUN	Air pressure	Riveting time	Stroke distance	T1S1	T1S2	T2S1	T2S2	Sum of squares of reciprocals	Mean of sum of squares of reciprocals	SN ratio (larger the better) N
1	0.3	1.5	2	0.10	1.00	0.10	0.50	205.000	51.250	-17.10
2	0.3	2	2.05	2.50	3.00	3.50	4.00	0.4152	0.1038	9.84
3	0.3	2.5	2.15	4.50	6.50	7.00	7.00	0.1139	0.0285	15.46
4	0.4	1.5	2	4.00	4.50	3.25	4.00	0.2691	0.0673	11.72
5	0.4	2	2.15	6.50	7.00	7.00	7.00	0.0849	0.0212	16.73
6	0.4	2.5	2	5.00	6.00	5.00	6.00	0.1356	0.0339	14.70
7	0.5	1.5	2.15	7.00	7.50	6.00	6.50	0.0896	0.0224	16.50
8	0.5	2	2	1.00	3.00	1.00	2.00	2.3611	0.5903	2.29
9	0.5	2.5	2.05	3.00	4.75	3.00	4.50	0.3159	0.0790	11.02

Figure 5 shows the main-effect graph, which indicates that factor C is the major possible contributor to the effect on the quality characteristics because the gradient of its line graph is the highest among all factors. However, this visual observation and this phenomenon have yet to be confirmed by ANOVA.

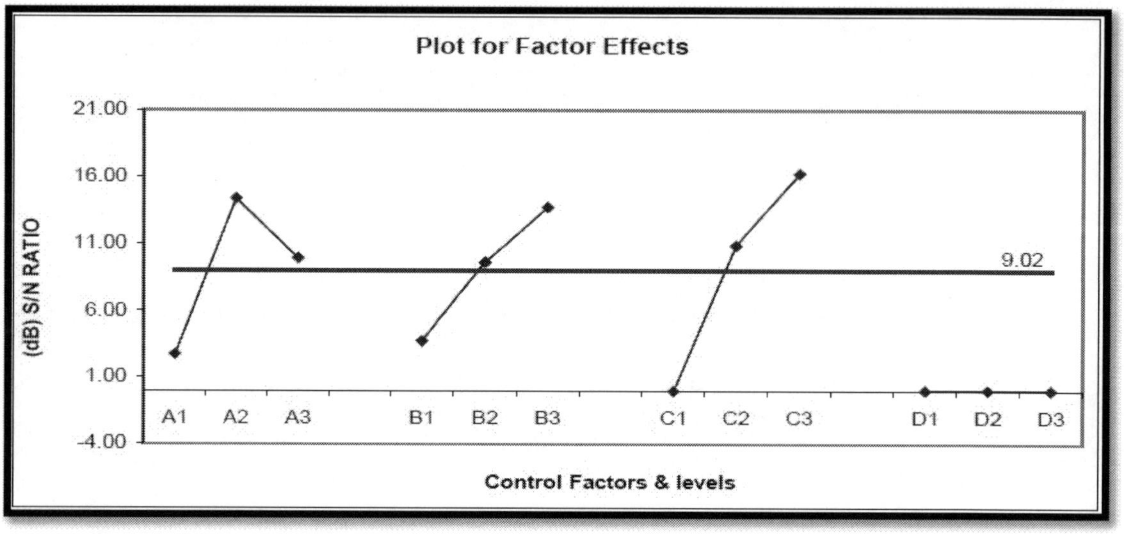

Figure 5. S/N ratio and main-effect plot for each factor. A and C represent the riveting strength experiment

Table 4 shows the ANOVA that calculates and estimates the effect of each factor and its contribution to quality performance. The ANOVA table reveals that factor C (stroke distance) contributes the most significant effect on the performance characteristic. The stroke-distance factor has the largest

F value at 2.63 and contributes 44.8%to the variation of the riveting strength. Factor A (air pressure) contributes 22.3% and an

F value of 1.32. The contribution of Factor B (riveting time) is not significant, and its

F value is 0.97; although its contribution is16.4%, the effect is very minimal to the performance characteristic. The

F values of factors A and B are less than 1.5; therefore, they are pooled for optimum response calculation.

	Factor/Source	Degree of freedom	Sum of Squares	Mean squares	Mean square ratio - F	Percent contribution %
	Table 4. The ANOVA for factor effect before pooling					
A	Air pressure	2	207.46	103.73	1.32	22.34
B	Riveting time	2	152.23	76.11	0.97	16.40
C	Stroke distance	2	412.08	206.04	2.63	44.38
	Error	2	156.67	78.3355		16.87
	Total	8	928.43			100

The experiment has resulted in an optimum S/N ratio value of 16.2 db when only the effect of factor A at level2 was taken into consideration. The preliminary optimum combination of the parameter setting was set as A2, B3, and B4, selected as suggested by the main effect plots (Figure 5). As the experiment had two response variables, the final setting was confirmed after the effect of the combination setting on the second response variables was taken into consideration. From the optimum S/N ratio, the confidence interval was computed to be from 5.8 db to 26.5 db. The calculation was made for a 90% confidence level [2]. The overall mean of the experiment was 9.02 db. The second quality performance in this experiment was the snap height. The L9 array combination result was used to measure the snap height performance, which was measured first prior to the destructive strength test. The result is shown in Table 6, which indicates that the overall mean was 0.0725 mm.

Tabla 5. ANOVA after pooling factors A and B

	Factor/ Source	Degrees of freedom	Sum of Squares	Mean Squares	Mean Square Ratio - F	X= not in used or Pooled Variance	Pure Sum of Squares	Percent Contribution, %
A	Air Pressure					X		
B	Riveting Time					X		
C	Stroke distance	**2**	**412.08**	**206.04**	**2.39**		**239.96**	**25.846**
	Error	**6**	**516.35**	**86.0588**			**688.470**	**74.154**
	Total	**8**	**928.43**					**100.00**

Table 6. The L9 matrix and the experimental results

	FACTORS			NOISE FACTOR						
RUN	Air pressure	Riveting time	Stroke Distance	T1S1	T1S2	T2S1	T2S2	Mean	Variance	SN Ratio (Nominal-the-Best)
1	0.3	1.5	2	0.15	0.14	0.14	0.15	0.15	0.00	31.12651161
2	0.3	2	2.05	0.10	0.10	0.10	0.11	0.10	0.00	27.613889
3	0.3	2.5	2.15	0.03	0.03	0.02	0.02	0.02	0.00	11.34742887
4	0.4	1.5	2.05	0.09	0.09	0.09	0.09	0.09	0.00	45.31992741
5	0.4	2	2.15	0.02	0.02	0.03	0.01	0.02	0.00	6.680725715
6	0.4	2.5	2	0.06	0.06	0.05	0.05	0.05	0.00	22.82357725
7	0.5	1.5	2.15	0.01	0.01	0.01	0.01	0.01	0.00	17.87893215
8	0.5	2	2	0.12	0.12	0.14	0.11	0.12	0.00	22.2012404
9	0.5	2.5	2.05	0.08	0.08	0.09	0.09	0.09	0.00	31.31839123

Figure 6 shows the main-effect plots, which indicate that factor C (the stroke distance) contributed the largest effect to the snap height, followed by factor B (the riveting time). Factor A (the air pressure) has shown no significant effect to the snap height. The important role of factor C is to ensure that sufficient force is applied to the riveting guide shaft, and factor B ensures that adequate time is allotted for completion of the riveting process before the radial riveting starts to return to its origin.

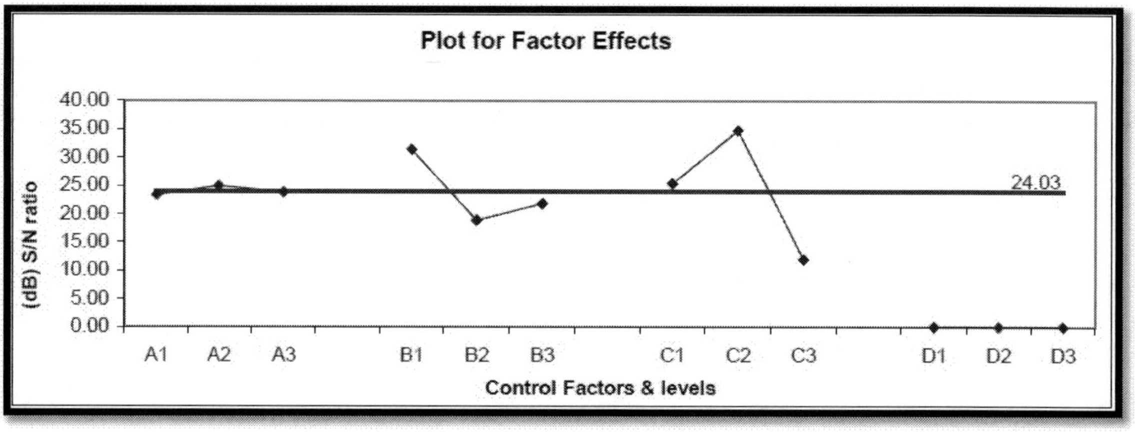

Figure 6. Main-effect plot of each factor level for the snap height

The mean effect of the factors was analyzed using the ANOVA method, and the result verified that factor C (the stroke distance) contributed 73.25% of the variation in snap height, with an F value calculated as 34.22. Factor B was the second contributor at 24.2% with an 11.32 F value. The non-significant factor A contributed only0.37 % with an F value of 0.17. This result is shown in Table 7. To determine the optimum condition of the response variable, factor A was pooled.

Table 7. ANOVA table for snap height before pooling

	Factors	Degree of freedom	Sum of Squares	Mean squares	Mean square ratio - F	Percent contribution
A	Air pressure	2	3.99	1.99	0.17	0.37
B	Riveting time	2	260.39	130.19	11.32	24.24
C	Stroke Distance	2	786.70	393.35	34.22	73.25
	Error	2	22.99	11.4961		2.14
	Total	8	1074.07			100
	Effective Dof factors	6				

The pooling of factor A resulted in higher *F* values for both the stroke distance and riveting time. This result is shown in Table 8, where the pure sum of squares is calculated.

Table 8. ANOVA table for snap height after pooling factor A

	Factors	Degree of freedom	Sum of Squares	Mean squares	Mean square ratio - F	X = not in used or Pooled variance	Pure sum of square	Percent Contribution %
A	Air pressure					x		
B	Riveting time	2	260.39	130.19	19.30		246.90	22.987
C	Stroke Distance	2	786.70	393.35	58.32		773.21	71.989
	Error	4	26.98	6.7449			53.959	5.024
	Total	8	1074.07					100.00
	Effective Dof factors	4						

The optimum result from this experiment was calculated based on the significant factors at the selected levels that affect the quality performance of the snap height. The optimum S/N ratio value was 42.16 db. The confidence interval selected was from the lower confidence range, from 0.0598 µm to 0.0850 µm, because the width is large enough for more practical applications of testing for the mean. The process of eliminating the contribution of the insignificant factors and adjusting the contribution of the other factors is called pooling. In pooling, the sum of squares of the insignificant factor effect is combined with error of the experiment.

In this case study, when the two non-significant factors were excluded, the process has increased the mean square error that resulted in a smaller percent contribution value from the significant factors, which occurred after all the errors and their effects were quantified and taken into account in calculating the pure sum of squares. The value of this pure sum of squares represents the actual effect of the significant factor (factors), and this value could be visualized in terms of the percentage contribution amount to the output result. The ANOVA significant factor effect justifies and confirms the decision to determine the optimum result and the confidence interval where the possible estimated process variation shall fall.

The increase in error will present a higher degree of freedom of error, and this condition will influence the estimated range of the confidence interval. The higher the error is, the smaller is the range, and this condition is more precise than that at larger range because only the effect from the main significant factor (factors) is being focused on. The optimum results would be inaccurate, and the confidence interval would be too large and impractical if the insignificant factors were not excluded from the calculation of the optimum value (in S/N ratio).

The effect of this result can also be seen in the conversion of the S/N ratio into predicted scale values. Statistical inference commonly represents the natural and actual distribution of the data; when the F value is small, this method excludes those factors to capture the amount of total error to be considered when estimating a pure relationship between the significant factor (factors) and its (their) effect. Verification is necessary to justify and validate the experiment. As the snap height is not a critical response variable, emphasis was concentrated more on ensuring that the riveting strength is maintained to attain better results. However, both quality characteristics confirmed that factor C (the stroke distance) is the main contributor to the variation effect. Ten samples were run under recommended combination settings: factor A at A2, 0.4 MPa; factor Bat B3, 2.5 sec; and factor C at C3, 2.15 mm.

The experiment result was found to be positive when the S/N ratio value for the riveting strength was 16.41 db, where it lay within the range of the calculated confidence interval. The S/N ratio result of the snap height was 28.9 db, better than the 12.06 db value before the experiment. The snap height was not tested with the optimum combination setting because the level set in the final parameter setting was different from the experiment. Priority was focused on ensuring the riveting strength as the main objective.

Comparison of the S/N ratio values was done to understand the achieved experimental result. The first condition was the value of S/N with the setting before the experiment, the optimum condition of the S/N ratio value from the experiment, and the value of the S/N ratio from the confirmatory run. The improvement in riveting strength achieved for the optimum condition in the actual verification run was 39% and 40%.

The "nominal-the-best" criterion for snap height was expected to be 250% at the optimum condition; however, its actual improvement following the riveting strength combination setting was only 134%, as explained and shown in Tables 9 and 10.

Table 9. Results and improvement of the experiment for snap height

	Control factors	Initial stage Initial condition	Optimum Predicted	Actual Observed
Parameter Setting	A ,B, C	A=0.4Mpa B= 1.5sec, C=2.10mm	A2 B1 C2	A2 B3 C3
S/N ratio (db)		12.06	42.16	28.19
% Improvement S/N ratio			250%	134%
Cpk against expected spec.		-0.075	-	3.512

Table 10. Improvement result of the experiment for critical riveting strength

	Control factors	Initial stage Initial Condition	Optimum Predicted	Confirmation Observed
Parameter Setting	A ,B, C	A=0.4Mpa B= 1.5sec, C=2.10mm	A2 B3 C3	A2 B3 C3
S/N ratio (db)		11.69	16.23	16.41
Riveting strength (Ncm)		3.84	6.48	6.61
% Improvement S/N ratio			39%	40%
% Improvement in scale unit			69%	72%

4. Conclusions

The Taguchi approach to DOEs is an effective strategy for product and process optimization. In this study, the Taguchi method has been proven to be a practical method to determine the optimum result through optimum combination parameter setting. The Taguchi method approach improved the process performance when optimum parameter settings used in the process resulted in better output performance in terms of quality and productivity. From this study, conclusion was made that the Taguchi method is an effective method in determining the optimum parameter setting and, finally, to quantify the achieved improvement. These results were observed from two experimental results. The actual improvement of the S/N ratio in riveting strength was 40%.

The Taguchi method has successfully determined the stroke distance as a significant factor that affected the performance of the riveting strength, technically proving that stroke distance is important in ensuring that sufficient displacement force is applied on the riveting head during the radial riveting process. If the distance is not enough, the amount of force is less, and the riveting process will be incomplete. From the same experiments, the stroke distance was found to be critical in ensuring that the snap height is maintained below 0.11 mm. This result proffered the conclusion that stroke distance is a critical factor because changes in this factor will cause a significant effect in the variation of the two quality performance criteria. The sensitivity of this stroke distance has triggered greater precaution in monitoring and control of this parameter.

References

1. Nalbant, M., Gokkaya, H., Sur, G., 2006. Application of Taguchi method in the optimization of cuttingparameters for surface roughness in turning. Materials and Design 28, 1379-1385.

2. Roy, R.K., 1990. A Primer on the Taguchi Method. Competitive Manufacturing Series. Van NosrrandReinhold, New York.

3. Phadke, M.S., 2008. Quality Engineering Using Robust Design. Low Price Edition. India: Pearson Edition,Inc. and Dorling Kindersley Publishing, Inc.

4. Dowey, S.J., Matthews, A., 1998. Taguchi and TQM Quality issues for surface engineered applications.Surface and Coating Technology 110, 86-93.

5.Ghani, J.A.,Choudhury, I.A.,Hassan, H.H.,2004. Application of Taguchi method in the optimization of end milling parameters.Journal of Materials Processing Technology145 (1), 84-92.

6. Apte, P.R., 2010. A 3-Day Course on Taguchi Method for Process and Product Optimization. IEEE-EDSMalaysia Chapter, Puri Pujangga, UKM, Bangi, Malaysia. 6-8 July 2010.

7. Zhang, J.Z., Chen, J.C., Kirby, E.D., 2006. Surface roughness optimization in an end-milling operation using the Taguchi design method. Journal of Materials Processing Technology 184, 233-239.

Appendix

t Table

cum. prob	$t_{.50}$	$t_{.75}$	$t_{.80}$	$t_{.85}$	$t_{.90}$	$t_{.95}$	$t_{.975}$	$t_{.99}$	$t_{.995}$	$t_{.999}$	$t_{.9995}$
one-tail	**0.50**	**0.25**	**0.20**	**0.15**	**0.10**	**0.05**	**0.025**	**0.01**	**0.005**	**0.001**	**0.0005**
two-tails	**1.00**	**0.50**	**0.40**	**0.30**	**0.20**	**0.10**	**0.05**	**0.02**	**0.01**	**0.002**	**0.001**
df											
1	0.000	1.000	1.376	1.963	3.078	6.314	12.71	31.82	63.66	318.31	636.62
2	0.000	0.816	1.061	1.386	1.886	2.920	4.303	6.965	9.925	22.327	31.599
3	0.000	0.765	0.978	1.250	1.638	2.353	3.182	4.541	5.841	10.215	12.924
4	0.000	0.741	0.941	1.190	1.533	2.132	2.776	3.747	4.604	7.173	8.610
5	0.000	0.727	0.920	1.156	1.476	2.015	2.571	3.365	4.032	5.893	6.869
6	0.000	0.718	0.906	1.134	1.440	1.943	2.447	3.143	3.707	5.208	5.959
7	0.000	0.711	0.896	1.119	1.415	1.895	2.365	2.998	3.499	4.785	5.408
8	0.000	0.706	0.889	1.108	1.397	1.860	2.306	2.896	3.355	4.501	5.041
9	0.000	0.703	0.883	1.100	1.383	1.833	2.262	2.821	3.250	4.297	4.781
10	0.000	0.700	0.879	1.093	1.372	1.812	2.228	2.764	3.169	4.144	4.587
11	0.000	0.697	0.876	1.088	1.363	1.796	2.201	2.718	3.106	4.025	4.437
12	0.000	0.695	0.873	1.083	1.356	1.782	2.179	2.681	3.055	3.930	4.318
13	0.000	0.694	0.870	1.079	1.350	1.771	2.160	2.650	3.012	3.852	4.221
14	0.000	0.692	0.868	1.076	1.345	1.761	2.145	2.624	2.977	3.787	4.140
15	0.000	0.691	0.866	1.074	1.341	1.753	2.131	2.602	2.947	3.733	4.073
16	0.000	0.690	0.865	1.071	1.337	1.746	2.120	2.583	2.921	3.686	4.015
17	0.000	0.689	0.863	1.069	1.333	1.740	2.110	2.567	2.898	3.646	3.965
18	0.000	0.688	0.862	1.067	1.330	1.734	2.101	2.552	2.878	3.610	3.922
19	0.000	0.688	0.861	1.066	1.328	1.729	2.093	2.539	2.861	3.579	3.883
20	0.000	0.687	0.860	1.064	1.325	1.725	2.086	2.528	2.845	3.552	3.850
21	0.000	0.686	0.859	1.063	1.323	1.721	2.080	2.518	2.831	3.527	3.819
22	0.000	0.686	0.858	1.061	1.321	1.717	2.074	2.508	2.819	3.505	3.792
23	0.000	0.685	0.858	1.060	1.319	1.714	2.069	2.500	2.807	3.485	3.768
24	0.000	0.685	0.857	1.059	1.318	1.711	2.064	2.492	2.797	3.467	3.745
25	0.000	0.684	0.856	1.058	1.316	1.708	2.060	2.485	2.787	3.450	3.725
26	0.000	0.684	0.856	1.058	1.315	1.706	2.056	2.479	2.779	3.435	3.707
27	0.000	0.684	0.855	1.057	1.314	1.703	2.052	2.473	2.771	3.421	3.690
28	0.000	0.683	0.855	1.056	1.313	1.701	2.048	2.467	2.763	3.408	3.674
29	0.000	0.683	0.854	1.055	1.311	1.699	2.045	2.462	2.756	3.396	3.659
30	0.000	0.683	0.854	1.055	1.310	1.697	2.042	2.457	2.750	3.385	3.646
40	0.000	0.681	0.851	1.050	1.303	1.684	2.021	2.423	2.704	3.307	3.551
60	0.000	0.679	0.848	1.045	1.296	1.671	2.000	2.390	2.660	3.232	3.460
80	0.000	0.678	0.846	1.043	1.292	1.664	1.990	2.374	2.639	3.195	3.416
100	0.000	0.677	0.845	1.042	1.290	1.660	1.984	2.364	2.626	3.174	3.390
1000	0.000	0.675	0.842	1.037	1.282	1.646	1.962	2.330	2.581	3.098	3.300
z	0.000	0.674	0.842	1.036	1.282	1.645	1.960	2.326	2.576	3.090	3.291
	0%	50%	60%	70%	80%	90%	95%	98%	99%	99.8%	99.9%
	Confidence Level										

Table entry for p is the critical value F^* with probability p lying to its right.

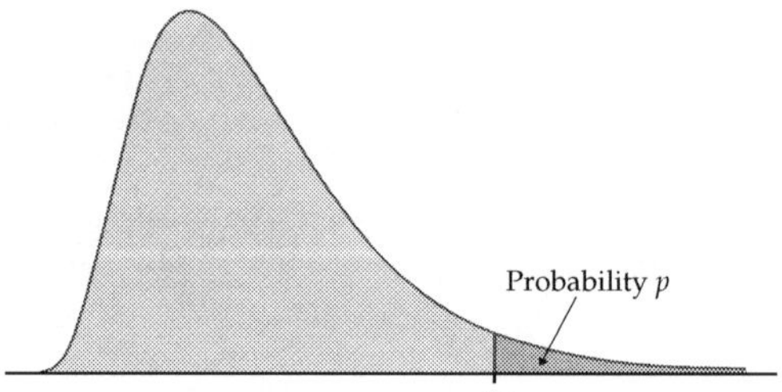

TABLE E
F critical values

Degrees of freedom in the denominator	p	Degrees of freedom in the numerator: 1	2	3	4	5	6	7	8	9
1	.100	39.86	49.50	53.59	55.83	57.24	58.20	58.91	59.44	59.86
	.050	161.45	199.50	215.71	224.58	230.16	233.99	236.77	238.88	240.54
	.025	647.79	799.50	864.16	899.58	921.85	937.11	948.22	956.66	963.28
	.010	4052.2	4999.5	5403.4	5624.6	5763.6	5859.0	5928.4	5981.1	6022.5
	.001	405284	500000	540379	562500	576405	585937	592873	598144	602284
2	.100	8.53	9.00	9.16	9.24	9.29	9.33	9.35	9.37	9.38
	.050	18.51	19.00	19.16	19.25	19.30	19.33	19.35	19.37	19.38
	.025	38.51	39.00	39.17	39.25	39.30	39.33	39.36	39.37	39.39
	.010	98.50	99.00	99.17	99.25	99.30	99.33	99.36	99.37	99.39
	.001	998.50	999.00	999.17	999.25	999.30	999.33	999.36	999.37	999.39
3	.100	5.54	5.46	5.39	5.34	5.31	5.28	5.27	5.25	5.24
	.050	10.13	9.55	9.28	9.12	9.01	8.94	8.89	8.85	8.81
	.025	17.44	16.04	15.44	15.10	14.88	14.73	14.62	14.54	14.47
	.010	34.12	30.82	29.46	28.71	28.24	27.91	27.67	27.49	27.35
	.001	167.03	148.50	141.11	137.10	134.58	132.85	131.58	130.62	129.86
4	.100	4.54	4.32	4.19	4.11	4.05	4.01	3.98	3.95	3.94
	.050	7.71	6.94	6.59	6.39	6.26	6.16	6.09	6.04	6.00
	.025	12.22	10.65	9.98	9.60	9.36	9.20	9.07	8.98	8.90
	.010	21.20	18.00	16.69	15.98	15.52	15.21	14.98	14.80	14.66
	.001	74.14	61.25	56.18	53.44	51.71	50.53	49.66	49.00	48.47
5	.100	4.06	3.78	3.62	3.52	3.45	3.40	3.37	3.34	3.32
	.050	6.61	5.79	5.41	5.19	5.05	4.95	4.88	4.82	4.77
	.025	10.01	8.43	7.76	7.39	7.15	6.98	6.85	6.76	6.68
	.010	16.26	13.27	12.06	11.39	10.97	10.67	10.46	10.29	10.16
	.001	47.18	37.12	33.20	31.09	29.75	28.83	28.16	27.65	27.24
6	.100	3.78	3.46	3.29	3.18	3.11	3.05	3.01	2.98	2.96
	.050	5.99	5.14	4.76	4.53	4.39	4.28	4.21	4.15	4.10
	.025	8.81	7.26	6.60	6.23	5.99	5.82	5.70	5.60	5.52
	.010	13.75	10.92	9.78	9.15	8.75	8.47	8.26	8.10	7.98
	.001	35.51	27.00	23.70	21.92	20.80	20.03	19.46	19.03	18.69
7	.100	3.59	3.26	3.07	2.96	2.88	2.83	2.78	2.75	2.72
	.050	5.59	4.74	4.35	4.12	3.97	3.87	3.79	3.73	3.68
	.025	8.07	6.54	5.89	5.52	5.29	5.12	4.99	4.90	4.82
	.010	12.25	9.55	8.45	7.85	7.46	7.19	6.99	6.84	6.72
	.001	29.25	21.69	18.77	17.20	16.21	15.52	15.02	14.63	14.33

Table entry for p is the critical value F^* with probability p lying to its right.

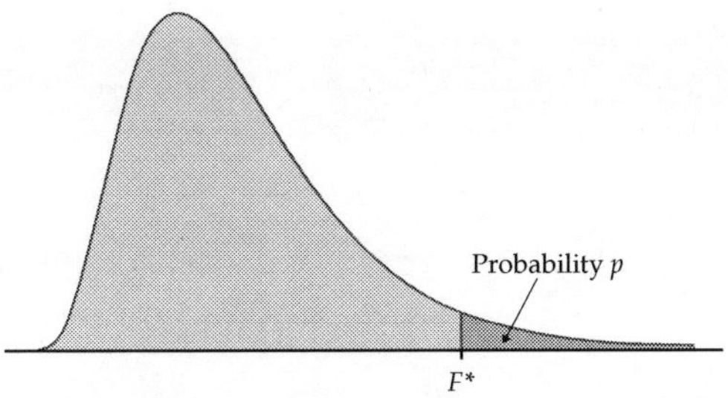

TABLE E

F critical values (continued)

Degrees of freedom in the numerator										
10	12	15	20	25	30	40	50	60	120	1000
60.19	60.71	61.22	61.74	62.05	62.26	62.53	62.69	62.79	63.06	63.30
241.88	243.91	245.95	248.01	249.26	250.10	251.14	251.77	252.20	253.25	254.19
968.63	976.71	984.87	993.10	998.08	1001.4	1005.6	1008.1	1009.8	1014.0	1017.7
6055.8	6106.3	6157.3	6208.7	6239.8	6260.6	6286.8	6302.5	6313.0	6339.4	6362.7
605621	610668	615764	620908	624017	626099	628712	630285	631337	633972	636301
9.39	9.41	9.42	9.44	9.45	9.46	9.47	9.47	9.47	9.48	9.49
19.40	19.41	19.43	19.45	19.46	19.46	19.47	19.48	19.48	19.49	19.49
39.40	39.41	39.43	39.45	39.46	39.46	39.47	39.48	39.48	39.49	39.50
99.40	99.42	99.43	99.45	99.46	99.47	99.47	99.48	99.48	99.49	99.50
999.40	999.42	999.43	999.45	999.46	999.47	999.47	999.48	999.48	999.49	999.50
5.23	5.22	5.20	5.18	5.17	5.17	5.16	5.15	5.15	5.14	5.13
8.79	8.74	8.70	8.66	8.63	8.62	8.59	8.58	8.57	8.55	8.53
14.42	14.34	14.25	14.17	14.12	14.08	14.04	14.01	13.99	13.95	13.91
27.23	27.05	26.87	26.69	26.58	26.50	26.41	26.35	26.32	26.22	26.14
129.25	128.32	127.37	126.42	125.84	125.45	124.96	124.66	124.47	123.97	123.53
3.92	3.90	3.87	3.84	3.83	3.82	3.80	3.80	3.79	3.78	3.76
5.96	5.91	5.86	5.80	5.77	5.75	5.72	5.70	5.69	5.66	5.63
8.84	8.75	8.66	8.56	8.50	8.46	8.41	8.38	8.36	8.31	8.26
14.55	14.37	14.20	14.02	13.91	13.84	13.75	13.69	13.65	13.56	13.47
48.05	47.41	46.76	46.10	45.70	45.43	45.09	44.88	44.75	44.40	44.09
3.30	3.27	3.24	3.21	3.19	3.17	3.16	3.15	3.14	3.12	3.11
4.74	4.68	4.62	4.56	4.52	4.50	4.46	4.44	4.43	4.40	4.37
6.62	6.52	6.43	6.33	6.27	6.23	6.18	6.14	6.12	6.07	6.02
10.05	9.89	9.72	9.55	9.45	9.38	9.29	9.24	9.20	9.11	9.03
26.92	26.42	25.91	25.39	25.08	24.87	24.60	24.44	24.33	24.06	23.82
2.94	2.90	2.87	2.84	2.81	2.80	2.78	2.77	2.76	2.74	2.72
4.06	4.00	3.94	3.87	3.83	3.81	3.77	3.75	3.74	3.70	3.67
5.46	5.37	5.27	5.17	5.11	5.07	5.01	4.98	4.96	4.90	4.86
7.87	7.72	7.56	7.40	7.30	7.23	7.14	7.09	7.06	6.97	6.89
18.41	17.99	17.56	17.12	16.85	16.67	16.44	16.31	16.21	15.98	15.77
2.70	2.67	2.63	2.59	2.57	2.56	2.54	2.52	2.51	2.49	2.47
3.64	3.57	3.51	3.44	3.40	3.38	3.34	3.32	3.30	3.27	3.23
4.76	4.67	4.57	4.47	4.40	4.36	4.31	4.28	4.25	4.20	4.15
6.62	6.47	6.31	6.16	6.06	5.99	5.91	5.86	5.82	5.74	5.66
14.08	13.71	13.32	12.93	12.69	12.53	12.33	12.20	12.12	11.91	11.72

(*Continued*)

TABLE E

F critical values (continued)

Degrees of freedom in the denominator	p	Degrees of freedom in the numerator: 1	2	3	4	5	6	7	8	9
8	.100	3.46	3.11	2.92	2.81	2.73	2.67	2.62	2.59	2.56
	.050	5.32	4.46	4.07	3.84	3.69	3.58	3.50	3.44	3.39
	.025	7.57	6.06	5.42	5.05	4.82	4.65	4.53	4.43	4.36
	.010	11.26	8.65	7.59	7.01	6.63	6.37	6.18	6.03	5.91
	.001	25.41	18.49	15.83	14.39	13.48	12.86	12.40	12.05	11.77
9	.100	3.36	3.01	2.81	2.69	2.61	2.55	2.51	2.47	2.44
	.050	5.12	4.26	3.86	3.63	3.48	3.37	3.29	3.23	3.18
	.025	7.21	5.71	5.08	4.72	4.48	4.32	4.20	4.10	4.03
	.010	10.56	8.02	6.99	6.42	6.06	5.80	5.61	5.47	5.35
	.001	22.86	16.39	13.90	12.56	11.71	11.13	10.70	10.37	10.11
10	.100	3.29	2.92	2.73	2.61	2.52	2.46	2.41	2.38	2.35
	.050	4.96	4.10	3.71	3.48	3.33	3.22	3.14	3.07	3.02
	.025	6.94	5.46	4.83	4.47	4.24	4.07	3.95	3.85	3.78
	.010	10.04	7.56	6.55	5.99	5.64	5.39	5.20	5.06	4.94
	.001	21.04	14.91	12.55	11.28	10.48	9.93	9.52	9.20	8.96
11	.100	3.23	2.86	2.66	2.54	2.45	2.39	2.34	2.30	2.27
	.050	4.84	3.98	3.59	3.36	3.20	3.09	3.01	2.95	2.90
	.025	6.72	5.26	4.63	4.28	4.04	3.88	3.76	3.66	3.59
	.010	9.65	7.21	6.22	5.67	5.32	5.07	4.89	4.74	4.63
	.001	19.69	13.81	11.56	10.35	9.58	9.05	8.66	8.35	8.12
12	.100	3.18	2.81	2.61	2.48	2.39	2.33	2.28	2.24	2.21
	.050	4.75	3.89	3.49	3.26	3.11	3.00	2.91	2.85	2.80
	.025	6.55	5.10	4.47	4.12	3.89	3.73	3.61	3.51	3.44
	.010	9.33	6.93	5.95	5.41	5.06	4.82	4.64	4.50	4.39
	.001	18.64	12.97	10.80	9.63	8.89	8.38	8.00	7.71	7.48
13	.100	3.14	2.76	2.56	2.43	2.35	2.28	2.23	2.20	2.16
	.050	4.67	3.81	3.41	3.18	3.03	2.92	2.83	2.77	2.71
	.025	6.41	4.97	4.35	4.00	3.77	3.60	3.48	3.39	3.31
	.010	9.07	6.70	5.74	5.21	4.86	4.62	4.44	4.30	4.19
	.001	17.82	12.31	10.21	9.07	8.35	7.86	7.49	7.21	6.98
14	.100	3.10	2.73	2.52	2.39	2.31	2.24	2.19	2.15	2.12
	.050	4.60	3.74	3.34	3.11	2.96	2.85	2.76	2.70	2.65
	.025	6.30	4.86	4.24	3.89	3.66	3.50	3.38	3.29	3.21
	.010	8.86	6.51	5.56	5.04	4.69	4.46	4.28	4.14	4.03
	.001	17.14	11.78	9.73	8.62	7.92	7.44	7.08	6.80	6.58
15	.100	3.07	2.70	2.49	2.36	2.27	2.21	2.16	2.12	2.09
	.050	4.54	3.68	3.29	3.06	2.90	2.79	2.71	2.64	2.59
	.025	6.20	4.77	4.15	3.80	3.58	3.41	3.29	3.20	3.12
	.010	8.68	6.36	5.42	4.89	4.56	4.32	4.14	4.00	3.89
	.001	16.59	11.34	9.34	8.25	7.57	7.09	6.74	6.47	6.26
16	.100	3.05	2.67	2.46	2.33	2.24	2.18	2.13	2.09	2.06
	.050	4.49	3.63	3.24	3.01	2.85	2.74	2.66	2.59	2.54
	.025	6.12	4.69	4.08	3.73	3.50	3.34	3.22	3.12	3.05
	.010	8.53	6.23	5.29	4.77	4.44	4.20	4.03	3.89	3.78
	.001	16.12	10.97	9.01	7.94	7.27	6.80	6.46	6.19	5.98
17	.100	3.03	2.64	2.44	2.31	2.22	2.15	2.10	2.06	2.03
	.050	4.45	3.59	3.20	2.96	2.81	2.70	2.61	2.55	2.49
	.025	6.04	4.62	4.01	3.66	3.44	3.28	3.16	3.06	2.98
	.010	8.40	6.11	5.19	4.67	4.34	4.10	3.93	3.79	3.68
	.001	15.72	10.66	8.73	7.68	7.02	6.56	6.22	5.96	5.75

TABLE E

F critical values (continued)

Degrees of freedom in the numerator										
10	12	15	20	25	30	40	50	60	120	1000
2.54	2.50	2.46	2.42	2.40	2.38	2.36	2.35	2.34	2.32	2.30
3.35	3.28	3.22	3.15	3.11	3.08	3.04	3.02	3.01	2.97	2.93
4.30	4.20	4.10	4.00	3.94	3.89	3.84	3.81	3.78	3.73	3.68
5.81	5.67	5.52	5.36	5.26	5.20	5.12	5.07	5.03	4.95	4.87
11.54	11.19	10.84	10.48	10.26	10.11	9.92	9.80	9.73	9.53	9.36
2.42	2.38	2.34	2.30	2.27	2.25	2.23	2.22	2.21	2.18	2.16
3.14	3.07	3.01	2.94	2.89	2.86	2.83	2.80	2.79	2.75	2.71
3.96	3.87	3.77	3.67	3.60	3.56	3.51	3.47	3.45	3.39	3.34
5.26	5.11	4.96	4.81	4.71	4.65	4.57	4.52	4.48	4.40	4.32
9.89	9.57	9.24	8.90	8.69	8.55	8.37	8.26	8.19	8.00	7.84
2.32	2.28	2.24	2.20	2.17	2.16	2.13	2.12	2.11	2.08	2.06
2.98	2.91	2.85	2.77	2.73	2.70	2.66	2.64	2.62	2.58	2.54
3.72	3.62	3.52	3.42	3.35	3.31	3.26	3.22	3.20	3.14	3.09
4.85	4.71	4.56	4.41	4.31	4.25	4.17	4.12	4.08	4.00	3.92
8.75	8.45	8.13	7.80	7.60	7.47	7.30	7.19	7.12	6.94	6.78
2.25	2.21	2.17	2.12	2.10	2.08	2.05	2.04	2.03	2.00	1.98
2.85	2.79	2.72	2.65	2.60	2.57	2.53	2.51	2.49	2.45	2.41
3.53	3.43	3.33	3.23	3.16	3.12	3.06	3.03	3.00	2.94	2.89
4.54	4.40	4.25	4.10	4.01	3.94	3.86	3.81	3.78	3.69	3.61
7.92	7.63	7.32	7.01	6.81	6.68	6.52	6.42	6.35	6.18	6.02
2.19	2.15	2.10	2.06	2.03	2.01	1.99	1.97	1.96	1.93	1.91
2.75	2.69	2.62	2.54	2.50	2.47	2.43	2.40	2.38	2.34	2.30
3.37	3.28	3.18	3.07	3.01	2.96	2.91	2.87	2.85	2.79	2.73
4.30	4.16	4.01	3.86	3.76	3.70	3.62	3.57	3.54	3.45	3.37
7.29	7.00	6.71	6.40	6.22	6.09	5.93	5.83	5.76	5.59	5.44
2.14	2.10	2.05	2.01	1.98	1.96	1.93	1.92	1.90	1.88	1.85
2.67	2.60	2.53	2.46	2.41	2.38	2.34	2.31	2.30	2.25	2.21
3.25	3.15	3.05	2.95	2.88	2.84	2.78	2.74	2.72	2.66	2.60
4.10	3.96	3.82	3.66	3.57	3.51	3.43	3.38	3.34	3.25	3.18
6.80	6.52	6.23	5.93	5.75	5.63	5.47	5.37	5.30	5.14	4.99
2.10	2.05	2.01	1.96	1.93	1.91	1.89	1.87	1.86	1.83	1.80
2.60	2.53	2.46	2.39	2.34	2.31	2.27	2.24	2.22	2.18	2.14
3.15	3.05	2.95	2.84	2.78	2.73	2.67	2.64	2.61	2.55	2.50
3.94	3.80	3.66	3.51	3.41	3.35	3.27	3.22	3.18	3.09	3.02
6.40	6.13	5.85	5.56	5.38	5.25	5.10	5.00	4.94	4.77	4.62
2.06	2.02	1.97	1.92	1.89	1.87	1.85	1.83	1.82	1.79	1.76
2.54	2.48	2.40	2.33	2.28	2.25	2.20	2.18	2.16	2.11	2.07
3.06	2.96	2.86	2.76	2.69	2.64	2.59	2.55	2.52	2.46	2.40
3.80	3.67	3.52	3.37	3.28	3.21	3.13	3.08	3.05	2.96	2.88
6.08	5.81	5.54	5.25	5.07	4.95	4.80	4.70	4.64	4.47	4.33
2.03	1.99	1.94	1.89	1.86	1.84	1.81	1.79	1.78	1.75	1.72
2.49	2.42	2.35	2.28	2.23	2.19	2.15	2.12	2.11	2.06	2.02
2.99	2.89	2.79	2.68	2.61	2.57	2.51	2.47	2.45	2.38	2.32
3.69	3.55	3.41	3.26	3.16	3.10	3.02	2.97	2.93	2.84	2.76
5.81	5.55	5.27	4.99	4.82	4.70	4.54	4.45	4.39	4.23	4.08
2.00	1.96	1.91	1.86	1.83	1.81	1.78	1.76	1.75	1.72	1.69
2.45	2.38	2.31	2.23	2.18	2.15	2.10	2.08	2.06	2.01	1.97
2.92	2.82	2.72	2.62	2.55	2.50	2.44	2.41	2.38	2.32	2.26
3.59	3.46	3.31	3.16	3.07	3.00	2.92	2.87	2.83	2.75	2.66
5.58	5.32	5.05	4.78	4.60	4.48	4.33	4.24	4.18	4.02	3.87

(*Continued*)

TABLE E

F critical values (continued)

Degrees of freedom in the denominator	*p*	1	2	3	4	5	6	7	8	9
		Degrees of freedom in the numerator								
18	.100	3.01	2.62	2.42	2.29	2.20	2.13	2.08	2.04	2.00
	.050	4.41	3.55	3.16	2.93	2.77	2.66	2.58	2.51	2.46
	.025	5.98	4.56	3.95	3.61	3.38	3.22	3.10	3.01	2.93
	.010	8.29	6.01	5.09	4.58	4.25	4.01	3.84	3.71	3.60
	.001	15.38	10.39	8.49	7.46	6.81	6.35	6.02	5.76	5.56
19	.100	2.99	2.61	2.40	2.27	2.18	2.11	2.06	2.02	1.98
	.050	4.38	3.52	3.13	2.90	2.74	2.63	2.54	2.48	2.42
	.025	5.92	4.51	3.90	3.56	3.33	3.17	3.05	2.96	2.88
	.010	8.18	5.93	5.01	4.50	4.17	3.94	3.77	3.63	3.52
	.001	15.08	10.16	8.28	7.27	6.62	6.18	5.85	5.59	5.39
20	.100	2.97	2.59	2.38	2.25	2.16	2.09	2.04	2.00	1.96
	.050	4.35	3.49	3.10	2.87	2.71	2.60	2.51	2.45	2.39
	.025	5.87	4.46	3.86	3.51	3.29	3.13	3.01	2.91	2.84
	.010	8.10	5.85	4.94	4.43	4.10	3.87	3.70	3.56	3.46
	.001	14.82	9.95	8.10	7.10	6.46	6.02	5.69	5.44	5.24
21	.100	2.96	2.57	2.36	2.23	2.14	2.08	2.02	1.98	1.95
	.050	4.32	3.47	3.07	2.84	2.68	2.57	2.49	2.42	2.37
	.025	5.83	4.42	3.82	3.48	3.25	3.09	2.97	2.87	2.80
	.010	8.02	5.78	4.87	4.37	4.04	3.81	3.64	3.51	3.40
	.001	14.59	9.77	7.94	6.95	6.32	5.88	5.56	5.31	5.11
22	.100	2.95	2.56	2.35	2.22	2.13	2.06	2.01	1.97	1.93
	.050	4.30	3.44	3.05	2.82	2.66	2.55	2.46	2.40	2.34
	.025	5.79	4.38	3.78	3.44	3.22	3.05	2.93	2.84	2.76
	.010	7.95	5.72	4.82	4.31	3.99	3.76	3.59	3.45	3.35
	.001	14.38	9.61	7.80	6.81	6.19	5.76	5.44	5.19	4.99
23	.100	2.94	2.55	2.34	2.21	2.11	2.05	1.99	1.95	1.92
	.050	4.28	3.42	3.03	2.80	2.64	2.53	2.44	2.37	2.32
	.025	5.75	4.35	3.75	3.41	3.18	3.02	2.90	2.81	2.73
	.010	7.88	5.66	4.76	4.26	3.94	3.71	3.54	3.41	3.30
	.001	14.20	9.47	7.67	6.70	6.08	5.65	5.33	5.09	4.89
24	.100	2.93	2.54	2.33	2.19	2.10	2.04	1.98	1.94	1.91
	.050	4.26	3.40	3.01	2.78	2.62	2.51	2.42	2.36	2.30
	.025	5.72	4.32	3.72	3.38	3.15	2.99	2.87	2.78	2.70
	.010	7.82	5.61	4.72	4.22	3.90	3.67	3.50	3.36	3.26
	.001	14.03	9.34	7.55	6.59	5.98	5.55	5.23	4.99	4.80
25	.100	2.92	2.53	2.32	2.18	2.09	2.02	1.97	1.93	1.89
	.050	4.24	3.39	2.99	2.76	2.60	2.49	2.40	2.34	2.28
	.025	5.69	4.29	3.69	3.35	3.13	2.97	2.85	2.75	2.68
	.010	7.77	5.57	4.68	4.18	3.85	3.63	3.46	3.32	3.22
	.001	13.88	9.22	7.45	6.49	5.89	5.46	5.15	4.91	4.71
26	.100	2.91	2.52	2.31	2.17	2.08	2.01	1.96	1.92	1.88
	.050	4.23	3.37	2.98	2.74	2.59	2.47	2.39	2.32	2.27
	.025	5.66	4.27	3.67	3.33	3.10	2.94	2.82	2.73	2.65
	.010	7.72	5.53	4.64	4.14	3.82	3.59	3.42	3.29	3.18
	.001	13.74	9.12	7.36	6.41	5.80	5.38	5.07	4.83	4.64
27	.100	2.90	2.51	2.30	2.17	2.07	2.00	1.95	1.91	1.87
	.050	4.21	3.35	2.96	2.73	2.57	2.46	2.37	2.31	2.25
	.025	5.63	4.24	3.65	3.31	3.08	2.92	2.80	2.71	2.63
	.010	7.68	5.49	4.60	4.11	3.78	3.56	3.39	3.26	3.15
	.001	13.61	9.02	7.27	6.33	5.73	5.31	5.00	4.76	4.57

TABLE E

F critical values (continued)

Degrees of freedom in the numerator										
10	12	15	20	25	30	40	50	60	120	1000
1.98	1.93	1.89	1.84	1.80	1.78	1.75	1.74	1.72	1.69	1.66
2.41	2.34	2.27	2.19	2.14	2.11	2.06	2.04	2.02	1.97	1.92
2.87	2.77	2.67	2.56	2.49	2.44	2.38	2.35	2.32	2.26	2.20
3.51	3.37	3.23	3.08	2.98	2.92	2.84	2.78	2.75	2.66	2.58
5.39	5.13	4.87	4.59	4.42	4.30	4.15	4.06	4.00	3.84	3.69
1.96	1.91	1.86	1.81	1.78	1.76	1.73	1.71	1.70	1.67	1.64
2.38	2.31	2.23	2.16	2.11	2.07	2.03	2.00	1.98	1.93	1.88
2.82	2.72	2.62	2.51	2.44	2.39	2.33	2.30	2.27	2.20	2.14
3.43	3.30	3.15	3.00	2.91	2.84	2.76	2.71	2.67	2.58	2.50
5.22	4.97	4.70	4.43	4.26	4.14	3.99	3.90	3.84	3.68	3.53
1.94	1.89	1.84	1.79	1.76	1.74	1.71	1.69	1.68	1.64	1.61
2.35	2.28	2.20	2.12	2.07	2.04	1.99	1.97	1.95	1.90	1.85
2.77	2.68	2.57	2.46	2.40	2.35	2.29	2.25	2.22	2.16	2.09
3.37	3.23	3.09	2.94	2.84	2.78	2.69	2.64	2.61	2.52	2.43
5.08	4.82	4.56	4.29	4.12	4.00	3.86	3.77	3.70	3.54	3.40
1.92	1.87	1.83	1.78	1.74	1.72	1.69	1.67	1.66	1.62	1.59
2.32	2.25	2.18	2.10	2.05	2.01	1.96	1.94	1.92	1.87	1.82
2.73	2.64	2.53	2.42	2.36	2.31	2.25	2.21	2.18	2.11	2.05
3.31	3.17	3.03	2.88	2.79	2.72	2.64	2.58	2.55	2.46	2.37
4.95	4.70	4.44	4.17	4.00	3.88	3.74	3.64	3.58	3.42	3.28
1.90	1.86	1.81	1.76	1.73	1.70	1.67	1.65	1.64	1.60	1.57
2.30	2.23	2.15	2.07	2.02	1.98	1.94	1.91	1.89	1.84	1.79
2.70	2.60	2.50	2.39	2.32	2.27	2.21	2.17	2.14	2.08	2.01
3.26	3.12	2.98	2.83	2.73	2.67	2.58	2.53	2.50	2.40	2.32
4.83	4.58	4.33	4.06	3.89	3.78	3.63	3.54	3.48	3.32	3.17
1.89	1.84	1.80	1.74	1.71	1.69	1.66	1.64	1.62	1.59	1.55
2.27	2.20	2.13	2.05	2.00	1.96	1.91	1.88	1.86	1.81	1.76
2.67	2.57	2.47	2.36	2.29	2.24	2.18	2.14	2.11	2.04	1.98
3.21	3.07	2.93	2.78	2.69	2.62	2.54	2.48	2.45	2.35	2.27
4.73	4.48	4.23	3.96	3.79	3.68	3.53	3.44	3.38	3.22	3.08
1.88	1.83	1.78	1.73	1.70	1.67	1.64	1.62	1.61	1.57	1.54
2.25	2.18	2.11	2.03	1.97	1.94	1.89	1.86	1.84	1.79	1.74
2.64	2.54	2.44	2.33	2.26	2.21	2.15	2.11	2.08	2.01	1.94
3.17	3.03	2.89	2.74	2.64	2.58	2.49	2.44	2.40	2.31	2.22
4.64	4.39	4.14	3.87	3.71	3.59	3.45	3.36	3.29	3.14	2.99
1.87	1.82	1.77	1.72	1.68	1.66	1.63	1.61	1.59	1.56	1.52
2.24	2.16	2.09	2.01	1.96	1.92	1.87	1.84	1.82	1.77	1.72
2.61	2.51	2.41	2.30	2.23	2.18	2.12	2.08	2.05	1.98	1.91
3.13	2.99	2.85	2.70	2.60	2.54	2.45	2.40	2.36	2.27	2.18
4.56	4.31	4.06	3.79	3.63	3.52	3.37	3.28	3.22	3.06	2.91
1.86	1.81	1.76	1.71	1.67	1.65	1.61	1.59	1.58	1.54	1.51
2.22	2.15	2.07	1.99	1.94	1.90	1.85	1.82	1.80	1.75	1.70
2.59	2.49	2.39	2.28	2.21	2.16	2.09	2.05	2.03	1.95	1.89
3.09	2.96	2.81	2.66	2.57	2.50	2.42	2.36	2.33	2.23	2.14
4.48	4.24	3.99	3.72	3.56	3.44	3.30	3.21	3.15	2.99	2.84
1.85	1.80	1.75	1.70	1.66	1.64	1.60	1.58	1.57	1.53	1.50
2.20	2.13	2.06	1.97	1.92	1.88	1.84	1.81	1.79	1.73	1.68
2.57	2.47	2.36	2.25	2.18	2.13	2.07	2.03	2.00	1.93	1.86
3.06	2.93	2.78	2.63	2.54	2.47	2.38	2.33	2.29	2.20	2.11
4.41	4.17	3.92	3.66	3.49	3.38	3.23	3.14	3.08	2.92	2.78

(*Continued*)

TABLE E

F critical values (continued)

Degrees of freedom in the denominator		Degrees of freedom in the numerator								
	p	1	2	3	4	5	6	7	8	9
28	.100	2.89	2.50	2.29	2.16	2.06	2.00	1.94	1.90	1.87
	.050	4.20	3.34	2.95	2.71	2.56	2.45	2.36	2.29	2.24
	.025	5.61	4.22	3.63	3.29	3.06	2.90	2.78	2.69	2.61
	.010	7.64	5.45	4.57	4.07	3.75	3.53	3.36	3.23	3.12
	.001	13.50	8.93	7.19	6.25	5.66	5.24	4.93	4.69	4.50
29	.100	2.89	2.50	2.28	2.15	2.06	1.99	1.93	1.89	1.86
	.050	4.18	3.33	2.93	2.70	2.55	2.43	2.35	2.28	2.22
	.025	5.59	4.20	3.61	3.27	3.04	2.88	2.76	2.67	2.59
	.010	7.60	5.42	4.54	4.04	3.73	3.50	3.33	3.20	3.09
	.001	13.39	8.85	7.12	6.19	5.59	5.18	4.87	4.64	4.45
30	.100	2.88	2.49	2.28	2.14	2.05	1.98	1.93	1.88	1.85
	.050	4.17	3.32	2.92	2.69	2.53	2.42	2.33	2.27	2.21
	.025	5.57	4.18	3.59	3.25	3.03	2.87	2.75	2.65	2.57
	.010	7.56	5.39	4.51	4.02	3.70	3.47	3.30	3.17	3.07
	.001	13.29	8.77	7.05	6.12	5.53	5.12	4.82	4.58	4.39
40	.100	2.84	2.44	2.23	2.09	2.00	1.93	1.87	1.83	1.79
	.050	4.08	3.23	2.84	2.61	2.45	2.34	2.25	2.18	2.12
	.025	5.42	4.05	3.46	3.13	2.90	2.74	2.62	2.53	2.45
	.010	7.31	5.18	4.31	3.83	3.51	3.29	3.12	2.99	2.89
	.001	12.61	8.25	6.59	5.70	5.13	4.73	4.44	4.21	4.02
50	.100	2.81	2.41	2.20	2.06	1.97	1.90	1.84	1.80	1.76
	.050	4.03	3.18	2.79	2.56	2.40	2.29	2.20	2.13	2.07
	.025	5.34	3.97	3.39	3.05	2.83	2.67	2.55	2.46	2.38
	.010	7.17	5.06	4.20	3.72	3.41	3.19	3.02	2.89	2.78
	.001	12.22	7.96	6.34	5.46	4.90	4.51	4.22	4.00	3.82
60	.100	2.79	2.39	2.18	2.04	1.95	1.87	1.82	1.77	1.74
	.050	4.00	3.15	2.76	2.53	2.37	2.25	2.17	2.10	2.04
	.025	5.29	3.93	3.34	3.01	2.79	2.63	2.51	2.41	2.33
	.010	7.08	4.98	4.13	3.65	3.34	3.12	2.95	2.82	2.72
	.001	11.97	7.77	6.17	5.31	4.76	4.37	4.09	3.86	3.69
100	.100	2.76	2.36	2.14	2.00	1.91	1.83	1.78	1.73	1.69
	.050	3.94	3.09	2.70	2.46	2.31	2.19	2.10	2.03	1.97
	.025	5.18	3.83	3.25	2.92	2.70	2.54	2.42	2.32	2.24
	.010	6.90	4.82	3.98	3.51	3.21	2.99	2.82	2.69	2.59
	.001	11.50	7.41	5.86	5.02	4.48	4.11	3.83	3.61	3.44
200	.100	2.73	2.33	2.11	1.97	1.88	1.80	1.75	1.70	1.66
	.050	3.89	3.04	2.65	2.42	2.26	2.14	2.06	1.98	1.93
	.025	5.10	3.76	3.18	2.85	2.63	2.47	2.35	2.26	2.18
	.010	6.76	4.71	3.88	3.41	3.11	2.89	2.73	2.60	2.50
	.001	11.15	7.15	5.63	4.81	4.29	3.92	3.65	3.43	3.26
1000	.100	2.71	2.31	2.09	1.95	1.85	1.78	1.72	1.68	1.64
	.050	3.85	3.00	2.61	2.38	2.22	2.11	2.02	1.95	1.89
	.025	5.04	3.70	3.13	2.80	2.58	2.42	2.30	2.20	2.13
	.010	6.66	4.63	3.80	3.34	3.04	2.82	2.66	2.53	2.43
	.001	10.89	6.96	5.46	4.65	4.14	3.78	3.51	3.30	3.13

TABLE E

F critical values (continued)

Degrees of freedom in the numerator										
10	12	15	20	25	30	40	50	60	120	1000
1.84	1.79	1.74	1.69	1.65	1.63	1.59	1.57	1.56	1.52	1.48
2.19	2.12	2.04	1.96	1.91	1.87	1.82	1.79	1.77	1.71	1.66
2.55	2.45	2.34	2.23	2.16	2.11	2.05	2.01	1.98	1.91	1.84
3.03	2.90	2.75	2.60	2.51	2.44	2.35	2.30	2.26	2.17	2.08
4.35	4.11	3.86	3.60	3.43	3.32	3.18	3.09	3.02	2.86	2.72
1.83	1.78	1.73	1.68	1.64	1.62	1.58	1.56	1.55	1.51	1.47
2.18	2.10	2.03	1.94	1.89	1.85	1.81	1.77	1.75	1.70	1.65
2.53	2.43	2.32	2.21	2.14	2.09	2.03	1.99	1.96	1.89	1.82
3.00	2.87	2.73	2.57	2.48	2.41	2.33	2.27	2.23	2.14	2.05
4.29	4.05	3.80	3.54	3.38	3.27	3.12	3.03	2.97	2.81	2.66
1.82	1.77	1.72	1.67	1.63	1.61	1.57	1.55	1.54	1.50	1.46
2.16	2.09	2.01	1.93	1.88	1.84	1.79	1.76	1.74	1.68	1.63
2.51	2.41	2.31	2.20	2.12	2.07	2.01	1.97	1.94	1.87	1.80
2.98	2.84	2.70	2.55	2.45	2.39	2.30	2.25	2.21	2.11	2.02
4.24	4.00	3.75	3.49	3.33	3.22	3.07	2.98	2.92	2.76	2.61
1.76	1.71	1.66	1.61	1.57	1.54	1.51	1.48	1.47	1.42	1.38
2.08	2.00	1.92	1.84	1.78	1.74	1.69	1.66	1.64	1.58	1.52
2.39	2.29	2.18	2.07	1.99	1.94	1.88	1.83	1.80	1.72	1.65
2.80	2.66	2.52	2.37	2.27	2.20	2.11	2.06	2.02	1.92	1.82
3.87	3.64	3.40	3.14	2.98	2.87	2.73	2.64	2.57	2.41	2.25
1.73	1.68	1.63	1.57	1.53	1.50	1.46	1.44	1.42	1.38	1.33
2.03	1.95	1.87	1.78	1.73	1.69	1.63	1.60	1.58	1.51	1.45
2.32	2.22	2.11	1.99	1.92	1.87	1.80	1.75	1.72	1.64	1.56
2.70	2.56	2.42	2.27	2.17	2.10	2.01	1.95	1.91	1.80	1.70
3.67	3.44	3.20	2.95	2.79	2.68	2.53	2.44	2.38	2.21	2.05
1.71	1.66	1.60	1.54	1.50	1.48	1.44	1.41	1.40	1.35	1.30
1.99	1.92	1.84	1.75	1.69	1.65	1.59	1.56	1.53	1.47	1.40
2.27	2.17	2.06	1.94	1.87	1.82	1.74	1.70	1.67	1.58	1.49
2.63	2.50	2.35	2.20	2.10	2.03	1.94	1.88	1.84	1.73	1.62
3.54	3.32	3.08	2.83	2.67	2.55	2.41	2.32	2.25	2.08	1.92
1.66	1.61	1.56	1.49	1.45	1.42	1.38	1.35	1.34	1.28	1.22
1.93	1.85	1.77	1.68	1.62	1.57	1.52	1.48	1.45	1.38	1.30
2.18	2.08	1.97	1.85	1.77	1.71	1.64	1.59	1.56	1.46	1.36
2.50	2.37	2.22	2.07	1.97	1.89	1.80	1.74	1.69	1.57	1.45
3.30	3.07	2.84	2.59	2.43	2.32	2.17	2.08	2.01	1.83	1.64
1.63	1.58	1.52	1.46	1.41	1.38	1.34	1.31	1.29	1.23	1.16
1.88	1.80	1.72	1.62	1.56	1.52	1.46	1.41	1.39	1.30	1.21
2.11	2.01	1.90	1.78	1.70	1.64	1.56	1.51	1.47	1.37	1.25
2.41	2.27	2.13	1.97	1.87	1.79	1.69	1.63	1.58	1.45	1.30
3.12	2.90	2.67	2.42	2.26	2.15	2.00	1.90	1.83	1.64	1.43
1.61	1.55	1.49	1.43	1.38	1.35	1.30	1.27	1.25	1.18	1.08
1.84	1.76	1.68	1.58	1.52	1.47	1.41	1.36	1.33	1.24	1.11
2.06	1.96	1.85	1.72	1.64	1.58	1.50	1.45	1.41	1.29	1.13
2.34	2.20	2.06	1.90	1.79	1.72	1.61	1.54	1.50	1.35	1.16
2.99	2.77	2.54	2.30	2.14	2.02	1.87	1.77	1.69	1.49	1.22

Table A-8. Critical values (at 5% and 1% significance levels) for Duncan's Multiple Range Test.

	α = .05 Number of consecutive means (p) to be compared															
df	2	3	4	5	6	7	8	9	10	12	14	16	18	20	22	100
1	17.97	17.97	17.97	17.97	17.97	17.97	17.97	17.97	17.97	17.97	17.97	17.97	17.97	17.97	17.97	17.97
2	6.085	6.085	6.085	6.085	6.085	6.085	6.085	6.085	6.085	6.085	6.085	6.085	6.085	6.085	6.085	6.085
3	4.501	4.516	4.516	4.516	4.516	4.516	4.516	4.516	4.516	4.516	4.516	4.516	4.516	4.516	4.516	4.516
4	3.927	4.013	4.033	4.033	4.033	4.033	4.033	4.033	4.033	4.033	4.033	4.033	4.033	4.033	4.033	4.033
5	3.635	3.749	3.797	3.814	3.814	3.814	3.814	3.814	3.814	3.814	3.814	3.814	3.814	3.814	3.814	3.814
6	3.461	3.587	3.649	3.694	3.697	3.697	3.697	3.697	3.697	3.697	3.697	3.697	3.697	3.697	3.697	3.697
11	3.113	3.256	3.342	3.397	3.435	3.462	3.480	3.493	3.501	3.509	3.510	3.510	3.510	3.510	3.510	3.510
12	3.082	3.225	3.313	3.370	3.410	3.439	3.459	3.474	3.484	3.496	3.499	3.499	3.499	3.499	3.499	3.499
13	3.055	3.200	3.289	3.348	3.389	3.419	3.442	3.458	3.470	3.484	3.490	3.490	3.490	3.490	3.490	3.490
14	3.033	3.178	3.268	3.329	3.372	3.403	3.426	3.444	3.457	3.474	3.482	3.484	3.485	3.485	3.485	3.485
15	3.014	3.160	3.250	3.312	3.356	3.389	3.413	3.432	3.446	3.465	3.476	3.480	3.481	3.481	3.481	3.481
16	2.998	3.144	3.235	3.298	3.343	3.376	3.402	3.422	3.437	3.458	3.470	3.477	3.478	3.478	3.478	3.478
17	2.984	3.130	3.222	3.285	3.331	3.366	3.392	3.412	3.429	3.451	3.465	3.473	3.476	3.476	3.476	3.476
18	2.971	3.118	3.210	3.274	3.321	3.356	3.383	3.405	3.421	3.445	3.460	3.470	3.474	3.474	3.474	3.474
19	2.960	3.107	3.199	3.264	3.311	3.347	3.375	3.397	3.415	3.440	3.456	3.467	3.472	3.474	3.474	3.474
20	2.950	3.097	3.190	3.255	3.303	3.339	3.368	3.391	3.409	3.436	3.453	3.464	3.470	3.473	3.474	3.474
24	2.919	3.066	3.160	3.226	3.276	3.315	3.345	3.370	3.390	3.420	3.441	3.456	3.465	3.471	3.475	3.477
30	2.888	3.035	3.131	3.199	3.250	3.290	3.322	3.349	3.371	3.405	3.430	3.447	3.460	3.470	3.477	3.486
40	2.858	3.006	3.102	3.171	3.224	3.266	3.300	3.328	3.352	3.390	3.418	3.439	3.456	3.469	3.479	3.504
60	2.829	2.976	3.073	3.143	3.198	3.241	3.277	3.307	3.333	3.374	3.406	3.431	3.451	3.467	3.481	3.537
120	2.800	2.947	2.045	3.116	3.172	3.217	3.254	3.287	3.314	3.359	3.394	3.423	3.446	3.466	3.483	3.691
∞	2.772	2.918	3.017	3.089	3.146	3.193	3.232	3.265	3.294	3.343	3.382	3.414	3.442	3.466	3.486	3.735

df	α = .05 Number of consecutive means (p) to be compared															
	2	3	4	5	6	7	8	9	10	12	14	16	18	20	22	100
1	90.03	90.03	90.03	90.03	90.03	90.03	90.03	90.03	90.03	90.03	90.03	90.03	90.03	90.03	90.03	90.03
2	14.04	14.04	14.04	14.04	14.04	14.04	14.04	14.04	14.04	14.04	14.04	14.04	14.04	14.04	14.04	14.04
3	8.261	8.231	8.321	8.321	8.321	8.321	8.321	8.321	8.321	8.321	8.321	8.321	8.321	8.321	8.321	8.321
4	6.512	6.677	6.740	6.756	6.756	6.756	6.756	6.756	6.756	6.756	6.756	6.756	6.756	6.756	6.756	6.76
5	5.702	5.893	5.989	6.040	6.065	6.074	6.074	6.074	6.074	6.074	6.074	6.074	6.074	6.074	6.074	6.074
6	5.243	5.439	5.549	5.614	5.655	5.680	5.694	5.701	5.703	5.703	5.703	5.703	5.703	5.703	5.703	5.703
7	4.949	5.145	5.260	5.334	5.383	5.416	5.439	5.454	5.464	5.472	5.472	5.472	5.472	5.472	5.472	5.472
8	4.746	4.939	5.057	5.135	5.189	5.227	5.256	5.276	5.291	5.309	5.316	5.317	5.317	5.317	5.317	5.317
9	4.596	4.787	4.906	4.986	5.043	5.086	5.118	5.142	5.160	5.185	5.199	5.205	5.206	5.206	5.206	5.206
10	4.482	4.671	4.790	4.871	4.931	4.975	5.010	5.037	5.058	5.088	5.106	5.117	5.122	5.124	5.124	5.124
11	4.392	4.579	4.697	4.780	4.841	4.887	4.924	4.952	4.975	5.009	5.031	5.045	5.054	5.059	5.061	5.061
12	4.320	4.504	4.622	4.706	4.767	4.815	4.852	4.883	4.907	4.944	4.969	4.986	4.998	5.066	5.010	5.011
13	4.260	4.442	4.560	4.644	4.706	4.755	4.793	4.824	4.850	4.889	4.917	4.937	4.950	4.960	4.966	4.972
14	4.210	4.391	4.508	4.591	4.654	4.704	4.743	4.775	4.802	4.843	4.872	4.894	4.910	4.921	4.929	4.940
15	4.168	4.347	4.463	4.547	4.610	4.660	4.700	4.733	4.760	4.803	4.834	4.857	4.874	4.887	4.897	4.914
16	4.131	4.309	4.425	4.509	4.572	4.622	4.663	4.696	4.724	4.768	4.800	4.825	4.844	4.858	4.869	4.892
17	4.099	4.275	4.391	4.475	4.539	4.589	4.620	4.664	4.693	4.738	4.771	4.797	4.816	4.832	4.844	4.874
18	4.071	4.246	4.362	4.445	4.509	4.560	4.601	4.635	4.664	4.711	4.745	4.772	4.792	4.808	4.821	4.858
19	4.046	4.220	4.335	4.419	4.483	4.534	4.575	4.610	4.639	4.686	4.722	4.749	4.771	4.788	4.802	4.845
20	4.024	4.197	4.312	4.395	4.459	4.510	4.552	4.587	4.617	4.664	4.701	4.729	4.751	4.769	4.786	4.833
24	3.956	4.126	4.239	4.322	4.386	4.437	4.480	4.516	4.546	4.596	4.634	4.665	4.690	4.710	4.727	4.802
30	3.889	4.056	4.168	4.250	4.314	4.366	4.409	4.445	4.477	4.528	4.569	4.601	4.628	4.650	4.669	4.777
40	3.825	3.988	4.098	4.180	4.244	4.296	4.339	4.376	4.408	4.461	4.503	4.537	4.566	4.591	4.611	4.764
60	3.762	3.922	4.031	4.111	4.174	4.226	4.270	4.307	4.340	4.394	4.438	4.474	4.504	4.530	4.552	4.765
120	3.702	3.858	3.965	4.044	4.107	4.158	4.202	4.239	4.272	4.327	4.372	4.410	4.442	4.469	4.494	4.770
∞	3.643	3.796	3.900	3.978	4.040	4.091	4.135	4.172	4.205	4.261	4.307	4.345	4.379	4.408	4.434	4.776

Table A-9a. Critical values [t(α,k-1,df)] to compare control against each of k-1 other treatments in one-sided Dunnett's tests.

	α = .05 (k-1)								
df	1	2	3	4	5	6	7	8	9
5	2.02	2.44	2.68	2.85	2.98	3.08	3.16	3.24	3.30
6	1.94	2.34	2.56	2.71	2.83	2.92	3.00	3.07	3.12
7	1.89	2.27	2.48	3.62	2.73	2.82	2.89	2.95	3.01
8	1.86	2.22	2.42	2.55	2.66	2.74	2.81	2.87	2.92
9	1.83	2.18	2.37	2.50	2.60	2.68	2.75	2.81	2.86
10	1.81	2.15	2.34	2.47	2.56	2.64	2.70	2.76	2.81
11	1.80	2.13	2.31	2.44	2.53	2.60	2.67	2.72	2.77
12	1.78	2.11	2.29	2.41	2.50	2.58	2.64	2.69	2.74
13	1.77	2.09	2.27	2.39	2.48	2.55	2.61	2.65	2.71
14	1.76	2.08	2.25	2.37	2.46	2.53	2.59	2.64	2.69
16	1.75	2.06	2.23	2.34	2.43	2.50	2.56	2.61	2.65
18	1.73	2.04	2.21	2.32	2.41	2.48	2.53	2.58	2.62
20	1.72	2.03	2.19	2.30	2.39	2.46	2.51	2.56	2.60
30	1.70	1.99	2.15	2.25	2.33	2.40	2.45	2.50	2.54
60	1.67	1.95	2.10	2.21	2.28	2.35	2.39	2.44	2.48
120	1.66	1.93	2.08	2.18	2.26	2.32	2.37	2.41	2.45
∞	1.64	1.92	2.06	2.16	2.23	2.29	2.34	2.38	2.42

	α = .01 (k-1)								
df	1	2	3	4	5	6	7	8	9
5	3.37	3.90	4.21	4.43	4.60	4.73	4.85	4.94	5.03
6	3.14	3.61	3.88	4.07	4.21	4.33	4.43	4.51	4.59
6	3.00	3.42	3.66	3.83	3.96	4.07	4.15	4.23	4.30
8	2.90	3.29	3.51	3.67	3.79	3.88	3.96	4.03	4.09
9	2.82	3.19	3.40	3.55	3.66	3.75	3.82	3.89	3.94
10	2.76	3.11	3.31	3.45	3.56	3.64	3.71	3.78	3.83
11	2.72	3.06	3.25	3.38	3.48	3.56	3.63	3.69	3.74
12	2.68	3.01	3.19	3.32	3.42	3.50	3.56	3.62	3.67
13	2.65	2.97	3.15	3.27	3.37	3.44	3.51	3.56	3.61
14	2.62	2.94	3.11	3.23	3.32	3.40	3.46	3.51	3.56
16	2.58	2.88	3.05	3.17	3.26	3.33	3.39	3.44	3.48
18	2.55	2.84	3.01	3.12	3.21	3.27	3.33	3.38	3.42
20	2.53	2.81	2.97	3.08	3.17	3.23	3.29	3.34	3.38
30	2.46	2.72	2.87	2.97	3.05	3.11	3.16	3.21	3.24
60	2.39	2.64	2.78	2.87	2.94	3.00	3.04	3.08	3.12
120	2.36	2.60	2.73	2.82	2.89	2.94	2.99	3.03	3.06
∞	2.33	2.56	2.68	2.77	2.84	2.89	2.93	2.97	3.00

Table A-9b. Critical values [t(α, k-1, df) } to compare control against each of k-1 other treatments in two-sided Dunnett's tests.

	α = .05 (k-1)									
df	1	2	3	4	5	6	7	8	10	20
5	2.57	3.03	3.29	3.48	3.62	3.73	3.82	3.90	4.03	4.42
6	2.45	2.86	3.10	3.26	3.39	3.49	3.57	3.64	3.76	4.11
7	2.36	2.75	2.97	3.12	3.24	3.33	3.41	3.47	3.58	3.91
8	2.31	2.67	2.88	3.02	3.13	3.22	3.29	3.35	3.46	3.76
9	2.26	2.61	2.81	2.95	3.05	3.14	3.20	3.26	3.36	3.65
10	2.23	2.57	2.76	2.89	2.99	3.07	3.14	3.19	3.29	3.57
11	2.20	2.53	2.72	2.84	2.94	3.02	3.08	3.14	3.23	3.50
12	2.18	2.50	2.68	2.81	2.90	3.98	3.04	3.09	3.18	3.45
13	2.16	2.48	2.65	2.78	2.87	2.94	3.00	3.06	3.14	3.40
14	2.14	2.46	2.63	2.75	2.84	2.91	2.97	3.02	3.11	3.36
16	2.12	2.42	2.59	2.71	2.80	2.87	2.92	2.97	3.06	3.30
18	2.10	2.40	2.56	2.68	2.76	2.83	2.89	2.94	3.01	3.25
20	2.09	2.38	2.54	2.65	2.73	2.80	2.86	2.90	2.98	3.22
30	2.04	2.32	2.47	2.58	2.66	2.72	2.77	2.82	2.89	3.11
60	2.00	2.27	2.41	2.51	2.58	2.64	2.69	2.73	2.80	3.00
120	1.98	2.24	2.38	2.47	2.55	2.60	2.65	2.69	2.76	2.95
∞	1.96	2.21	2.35	2.44	2.51	2.57	2.61	2.65	2.72	2.91

	α = .01 (k-1)									
df	1	2	3	4	5	6	7	8	10	20
5	4.03	4.63	4.98	5.22	5.41	5.56	5.69	5.80	5.98	6.52
6	3.71	4.21	4.51	4.71	4.87	5.00	5.10	5.20	5.35	5.81
7	3.50	3.95	4.21	4.39	4.53	4.64	4.74	4.82	4.95	5.36
8	3.36	3.77	4.00	4.17	4.29	4.40	4.48	4.56	4.68	5.05
9	3.25	3.63	3.85	4.01	4.12	4.22	4.30	4.37	4.48	4.82
10	3.17	3.53	3.74	3.88	3.99	4.08	4.16	4.22	4.33	4.65
11	3.11	3.45	3.65	3.79	3.89	3.98	4.05	4.11	4.21	4.52
12	3.05	3.39	3.58	3.71	3.81	3.89	3.96	4.02	4.12	4.41
13	3.01	3.33	3.52	3.65	3.74	3.82	3.89	3.94	4.04	4.32
14	2.98	3.29	3.47	3.59	3.69	3.76	3.83	3.88	3.97	4.24
16	2.92	3.22	3.39	3.51	3.60	3.67	3.73	3.78	3.87	4.13
18	2.88	3.17	3.33	3.44	3.53	3.60	3.66	3.71	3.79	4.04
20	2.85	3.13	3.29	3.40	3.48	3.55	3.60	3.65	3.73	3.97
30	2.75	3.01	3.15	3.25	3.33	3.39	3.44	3.49	3.56	3.78
60	2.66	2.90	3.03	3.12	3.19	3.25	3.29	3.33	3.40	3.56
120	2.62	2.85	2.97	3.06	3.12	3.18	3.22	3.26	3.32	3.51
∞	2.58	2.79	2.92	3.00	3.06	3.11	3.15	3.19	3.25	3.42

Table A-10. Orthogonal coefficients for trend comparisons.

k	polynomial	coefficients										Σc_{ij}^2
3	Linear	-1	0	1								2
	Quadratic	1	-2	1								6
4	Linear	-3	-1	1	3							20
	Quadratic	1	-1	-1	1							4
	Cubic	-1	3	-3	1							20
5	Linear	-2	-1	0	1	2						10
	Quadratic	2	-1	-2	-1	2						14
	Cubic	-1	2	0	-2	1						10
	Quartic	1	-4	6	-4	1						70
6	Linear	-5	-3	-1	1	3	5					70
	Quadratic	5	-1	-4	-4	-1	5					84
	Cubic	-5	7	4	-4	-7	5					180
	Quartic	1	-3	2	2	-3	1					28
7	Linear	-3	-2	-1	0	1	2	3				28
	Quadratic	5	0	-3	-4	-3	0	5				84
	Cubic	-1	1	1	0	-1	-1	1				6
	Quartic	3	-7	1	6	1	-7	3				154
8	Linear	-7	-5	-3	-1	1	3	5	7			168
	Quadratic	7	1	-3	-5	-5	-3	1	7			168
	Cubic	-7	5	7	3	3	-7	-5	7			264
	Quartic	7	-13	-3	9	9	-3	-13	7			616
	Quintic	-7	23	-17	-15	15	17	-23	7			2184
9	Linear	-4	-3	-2	-1	0	1	2	3	4		60
	Quadratic	28	7	-8	-17	-20	-17	-8	7	28		2772
	Cubic	-14	7	13	9	0	-9	-13	-7	14		990
	Quartic	14	-21	-11	9	18	9	-11	-21	14		2002
	Quintic	-4	11	-4	-9	0	9	4	-11	4		468
10	Linear	-9	-7	-5	-3	-1	1	3	5	7	9	330
	Quadratic	6	2	-1	-3	-4	-4	-3	-1	2	6	132
	Cubic	-42	14	35	31	12	-12	-31	-35	-14	42	8580
	Quartic	18	-22	-17	3	18	18	3	-17	-22	18	2860
	Quintic	-6	14	-1	-11	-6	6	11	1	-14	6	780

Table A-11. Critical values of the correlation coefficient for certain levels (.1, .05, .01, .001)of significance.

df	.1	.05	.01	.001
1	.9879	.9969	.9999	1.0000
2	.9000	.9500	.9900	.9990
3	.8054	.8783	.9587	.9912
4	.7293	.8114	.9172	9741
5	.6694	.7545	.8745	9507
6	.6215	.7067	.8343	.9249
7	.5822	.6664	.7977	. 8082
8	.5494	.6319	.7646	. 8721
9	.5214	.6021	.7348	. 8471
10	.4973	.5760	.7079	. 8233
11	.4762	.5529	.6835	. 8010
12	.4575	.5324	.6614	. 7800
13	.4409	.5139	.6411	. 7603
14	.4259	.4973	.6226	. 7420
15	.4124	.4821	.6055	. 7246
16	.4000	.4683	.5897	. 7084
17	.3887	.4555	.5751	. 6932
18	.3783	.4439	.5614	. 6787
19	.3687	.4329	.5487	. 6653
20	.3598	.4227	.5368	. 6524
25	.3233	.38-9	.4869	. 5974
30	.2960	.3494	.4487	. 5541
35	.2746	.3246	.4182	. 5189
40	.2573	.3044	.3932	. 4896
45	.2428	.2875	.3721	. 4648
50	.2306	.2732	.3541	. 4433
60	.2108	.2500	.3248	. 4078
70	.1954	.2319	.3017	. 3799
80	.1829	.2172	.2830	. 3568
90	.1726	.2050	.2673	. 3375
100	.1638	.1946	.2540	. 3211

TABLE: Q SCORES FOR TUKEY'S METHOD

	$\alpha = 0.05$										$\alpha = 0.01$								
k / df	2	3	4	5	6	7	8	9	10	k / df	2	3	4	5	6	7	8	9	10
1	18.0	27.0	32.8	37.1	40.4	43.1	45.4	47.4	49.1	1	90.0	135	164	186	202	216	227	237	246
2	6.08	8.33	9.80	10.88	11.73	12.43	13.03	13.54	13.99	2	13.90	19.02	22.56	25.37	27.76	29.86	31.73	33.41	34.93
3	4.50	5.91	6.82	7.50	8.04	8.48	8.85	9.18	9.46	3	8.26	10.62	12.17	13.32	14.24	15.00	15.65	16.21	16.71
4	3.93	5.04	5.76	6.29	6.71	7.05	7.35	7.60	7.83	4	6.51	8.12	9.17	9.96	10.58	11.10	11.54	11.92	12.26
5	3.64	4.60	5.22	5.67	6.03	6.33	6.58	6.80	6.99	5	5.70	6.98	7.80	8.42	8.91	9.32	9.67	9.97	10.24
6	3.46	4.34	4.90	5.30	5.63	5.90	6.12	6.32	6.49	6	5.24	6.33	7.03	7.56	7.97	8.32	8.61	8.87	9.10
7	3.34	4.16	4.68	5.06	5.36	5.61	5.82	6.00	6.16	7	4.95	5.92	6.54	7.00	7.37	7.68	7.94	8.17	8.37
8	3.26	4.04	4.53	4.89	5.17	5.40	5.60	5.77	5.92	8	4.75	5.64	6.20	6.62	6.96	7.24	7.47	7.68	7.86
9	3.20	3.95	4.41	4.76	5.02	5.24	5.43	5.59	5.74	9	4.60	5.43	5.96	6.35	6.66	6.91	7.13	7.33	7.49
10	3.15	3.88	4.33	4.65	4.91	5.12	5.30	5.46	5.60	10	4.48	5.27	5.77	6.14	6.43	6.67	6.87	7.05	7.21
11	3.11	3.82	4.26	4.57	4.82	5.03	5.20	5.35	5.49	11	4.39	5.15	5.62	5.97	6.25	6.48	6.67	6.84	6.99
12	3.08	3.77	4.20	4.51	4.75	4.95	5.12	5.27	5.39	12	4.32	5.05	5.50	5.84	6.10	6.32	6.51	6.67	6.81
13	3.06	3.73	4.15	4.45	4.69	4.88	5.05	5.19	5.32	13	4.26	4.96	5.40	5.73	5.98	6.19	6.37	6.53	6.67
14	3.03	3.70	4.11	4.41	4.64	4.83	4.99	5.13	5.25	14	4.21	4.89	5.32	5.63	5.88	6.08	6.26	6.41	6.54
15	3.01	3.67	4.08	4.37	4.59	4.78	4.94	5.08	5.20	15	4.17	4.84	5.25	5.56	5.80	5.99	6.16	6.31	6.44
16	3.00	3.65	4.05	4.33	4.56	4.74	4.90	5.03	5.15	16	4.13	4.79	5.19	5.49	5.72	5.92	6.08	6.22	6.35
17	2.98	3.63	4.02	4.30	4.52	4.70	4.86	4.99	5.11	17	4.10	4.74	5.14	5.43	5.66	5.85	6.01	6.15	6.27
18	2.97	3.61	4.00	4.28	4.49	4.67	4.82	4.96	5.07	18	4.07	4.70	5.09	5.38	5.60	5.79	5.94	6.08	6.20
19	2.96	3.59	3.98	4.25	4.47	4.65	4.79	4.92	5.04	19	4.05	4.67	5.05	5.33	5.55	5.73	5.89	6.02	6.14
20	2.95	3.58	3.96	4.23	4.45	4.62	4.77	4.90	5.01	20	4.02	4.64	5.02	5.29	5.51	5.69	5.84	5.97	6.09
24	2.92	3.53	3.90	4.17	4.37	4.54	4.68	4.81	4.92	24	3.96	4.55	4.91	5.17	5.37	5.54	5.69	5.81	5.92
30	2.89	3.49	3.85	4.10	4.30	4.46	4.60	4.72	4.82	30	3.89	4.45	4.80	5.05	5.24	5.40	5.54	5.65	5.76
40	2.86	3.44	3.79	4.04	4.23	4.39	4.52	4.63	4.73	40	3.82	4.37	4.70	4.93	5.11	5.26	5.39	5.50	5.60
60	2.83	3.40	3.74	3.98	4.16	4.31	4.44	4.55	4.65	60	3.76	4.28	4.59	4.82	4.99	5.13	5.25	5.36	5.45
120	2.80	3.36	3.68	3.92	4.10	4.24	4.36	4.47	4.56	120	3.70	4.20	4.50	4.71	4.87	5.01	5.12	5.21	5.30
∞	2.77	3.31	3.63	3.86	4.03	4.17	4.29	4.39	4.47	∞	3.64	4.12	4.40	4.60	4.76	4.88	4.99	5.08	5.16

Chi-square Distribution Table

d.f.	.995	.99	.975	.95	.9	.1	.05	.025	.01
1	0.00	0.00	0.00	0.00	0.02	2.71	3.84	5.02	6.63
2	0.01	0.02	0.05	0.10	0.21	4.61	5.99	7.38	9.21
3	0.07	0.11	0.22	0.35	0.58	6.25	7.81	9.35	11.34
4	0.21	0.30	0.48	0.71	1.06	7.78	9.49	11.14	13.28
5	0.41	0.55	0.83	1.15	1.61	9.24	11.07	12.83	15.09
6	0.68	0.87	1.24	1.64	2.20	10.64	12.59	14.45	16.81
7	0.99	1.24	1.69	2.17	2.83	12.02	14.07	16.01	18.48
8	1.34	1.65	2.18	2.73	3.49	13.36	15.51	17.53	20.09
9	1.73	2.09	2.70	3.33	4.17	14.68	16.92	19.02	21.67
10	2.16	2.56	3.25	3.94	4.87	15.99	18.31	20.48	23.21
11	2.60	3.05	3.82	4.57	5.58	17.28	19.68	21.92	24.72
12	3.07	3.57	4.40	5.23	6.30	18.55	21.03	23.34	26.22
13	3.57	4.11	5.01	5.89	7.04	19.81	22.36	24.74	27.69
14	4.07	4.66	5.63	6.57	7.79	21.06	23.68	26.12	29.14
15	4.60	5.23	6.26	7.26	8.55	22.31	25.00	27.49	30.58
16	5.14	5.81	6.91	7.96	9.31	23.54	26.30	28.85	32.00
17	5.70	6.41	7.56	8.67	10.09	24.77	27.59	30.19	33.41
18	6.26	7.01	8.23	9.39	10.86	25.99	28.87	31.53	34.81
19	6.84	7.63	8.91	10.12	11.65	27.20	30.14	32.85	36.19
20	7.43	8.26	9.59	10.85	12.44	28.41	31.41	34.17	37.57
22	8.64	9.54	10.98	12.34	14.04	30.81	33.92	36.78	40.29
24	9.89	10.86	12.40	13.85	15.66	33.20	36.42	39.36	42.98
26	11.16	12.20	13.84	15.38	17.29	35.56	38.89	41.92	45.64
28	12.46	13.56	15.31	16.93	18.94	37.92	41.34	44.46	48.28
30	13.79	14.95	16.79	18.49	20.60	40.26	43.77	46.98	50.89
32	15.13	16.36	18.29	20.07	22.27	42.58	46.19	49.48	53.49
34	16.50	17.79	19.81	21.66	23.95	44.90	48.60	51.97	56.06
38	19.29	20.69	22.88	24.88	27.34	49.51	53.38	56.90	61.16
42	22.14	23.65	26.00	28.14	30.77	54.09	58.12	61.78	66.21
46	25.04	26.66	29.16	31.44	34.22	58.64	62.83	66.62	71.20
50	27.99	29.71	32.36	34.76	37.69	63.17	67.50	71.42	76.15
55	31.73	33.57	36.40	38.96	42.06	68.80	73.31	77.38	82.29
60	35.53	37.48	40.48	43.19	46.46	74.40	79.08	83.30	88.38
65	39.38	41.44	44.60	47.45	50.88	79.97	84.82	89.18	94.42
70	43.28	45.44	48.76	51.74	55.33	85.53	90.53	95.02	100.43
75	47.21	49.48	52.94	56.05	59.79	91.06	96.22	100.84	106.39
80	51.17	53.54	57.15	60.39	64.28	96.58	101.88	106.63	112.33
85	55.17	57.63	61.39	64.75	68.78	102.08	107.52	112.39	118.24
90	59.20	61.75	65.65	69.13	73.29	107.57	113.15	118.14	124.12
95	63.25	65.90	69.92	73.52	77.82	113.04	118.75	123.86	129.97
100	67.33	70.06	74.22	77.93	82.36	118.50	124.34	129.56	135.81

Coefficients of Orthogonal Polynomials

Polynomial	X=1	2	3	4	5	6	7	8	9	10
Linear	-1	0	1							
Quadratic	1	-2	1							
Linear	-3	-1	1	3						
Quadratic	1	-1	-1	1						
Cubic	-1	3	-3	1						
Linear	-2	-1	0	1	2					
Quadratic	2	-1	-2	-1	2					
Cubic	-1	2	0	-2	1					
Quartic	1	-4	6	-4	1					
Linear	-5	-3	-1	1	3	5				
Quadratic	5	-1	-4	-4	-1	5				
Cubic	-5	7	4	-4	-7	5				
Quartic	1	-3	2	2	-3	1				
Linear	-3	-2	-1	0	1	2	3			
Quadratic	5	0	-3	-4	-3	0	5			
Cubic	-1	1	1	0	-1	-1	1			
Quartic	3	-7	1	6	1	-7	3			
Linear	-7	-5	-3	-1	1	3	5	7		
Quadratic	7	1	-3	-5	-5	-3	1	7		
Cubic	-7	5	7	3	-3	-7	-5	7		
Quartic	7	-13	-3	9	9	-3	-13	7		
Quintic	-7	23	-17	-15	15	17	-23	7		
Linear	-4	-3	-2	-1	0	1	2	3	4	
Quadratic	28	7	-8	-17	-20	-17	-8	7	28	
Cubic	-14	7	13	9	0	-9	-13	-7	14	
Quartic	14	-21	-11	9	18	9	-11	-21	14	
Quintic	-4	11	-4	-9	0	9	4	-11	4	

References:

Anderson, V.L. and R.A. McLean. Design of Experiments. A Realistic Approach. Marcel Dekker, Inc., New York. 1972.

Barlett, M.S. A Note on the Analysis of Covariance. Jour. Agric. Sci., 26, 488-491. 1936. Barlett, M.S. The Use of Transformations. Biometrics, 3, 39-52. 1947.

Cochran, W.G. and G.M. Cox. Experimental Designs. John Wiley & Sons, New York. 1950.

Edwards, A., Experimental Design in Psychological Research, 5th edition, New York, Holt, Rinehart and Winston, 1984.

Ferber, R., and W. Hirsch, Social Experimentation and Economic Policy, Cambridge, England, Cambridge University Press, 1982.

Finney, D.J. An Introduction to the Theory of Experimental Designs. University of Chicago Press. 1960.

Fisher, R.A. The Design of Experiments. Oliver and Boyd, Edinburgh. 1947.

Fisher, R., Statistical Methods for Research Workers, 14th edition, Edinburgh, Scotland, Oliver and Boyd 1970.

Fisher, R., The Design of Experiments, 9th edition, New York, Hafner, 1971.

Fleiss, J., The Design and Analysis of Clinical Experiments, New York, Wiley. (reprint of 1986 publication), 1999.

Girden, E., ANOVA: Repeated Measures, Newbury Park, CA, Sage Publications, 1992.

Glass, G., V. Willson, and J. Gottman, Design and Analysis of Time-Series Experiments, Revised edition, Charlotte, NC, Information Age Publishing, 2008.

Goupy, J., Methods for Experimental Design: Principles and Applications for Physicists and Chemists, Amsterdam, Elsevier Science, 1993.

Hicks, C.R. Fundamental Concepts in the Design of Experiments. Holt, Rinehart and Winston, New York. 1973.

Kempthorne, 0. Design and Analysis of Experiments. Robert E. Krieger Publishing Co., New York. 1979.

Kuehl, R., Statistical Principles for Research Design and Analysis, Belmont, CA, Duxbury Press, 1994.

Kuehl, R., Design of Experiments: Statistical Principles for Research Design and Analysis, 2nd

edition, Pacific Grove, CA, Duxbury/Thomson Learning, 2000.

Li, C., Introduction to Experimental Statistics, New York, McGraw-Hill, 1964.

Little, R., and D. Rubin, Statistical Analysis with Missing Data, 2nd edition, New York, Wiley, 2002.

Lucas, J. (1976), "Which Response Surface Design Is Best?" Technometrics, vol. 18, pp. 411–417.

Little, T.M. and F.J. Hills. Statistical Methods in Agricultural Research. Editorial Trillas, Mexico. 1983.

Mendenhall, W. Introduction to Probability and Statistics. Duxbury Press, Massachusetts. 1975.

Mason, R., Gunst, R., and Hess, J. Statistical Design and Analysis of Experiments: With Applications to Engineering and Science, 2nd edition, New York, Wiley, 2003.

McLean, R., and V. Anderson, Applied Factorial and Fractional Systems, New York, Marcel Dekker, 1984.

Montgomery, D., Design and Analysis of Experiments, 9th edition, New York, Wiley, 2017.

Montgomery, D., Peck, E., and Vining, G. Introduction to Linear Regression Analysis, 5th edition, New York, Wiley, 2012.

Myers, R. Classical and Modern Regression with Applications, 2nd edition, Pacific Grove, CA, Duxbury Press, 2000.

Myers, R., D. Montgomery, and C. Anderson-Cook, Response Surface Methodology: Process and Product Optimization Using Designed Experiments, 4th edition, Hoboken, NJ, Wiley, 2016.

Park, S., Robust Design for Quality Engineering, New York, Chapman & Hall, 1996.

Peng, K. C., The Design and Analysis of Scientific Experiments, Reading, MA, Addison Wesley, 1967.

Piepel, G., and J. Cornell (1994), "Mixture Experiment Approaches: Examples, Discussions, and Recommendations." Journal of Quality Technology, vol. 26, pp. 177–196.

Plackett, B., and J. Burman (1946), "The Design of Optimum Multifactorial Experiments." Biometrika, vol. 33, pp. 305–325.

Owen, D.B. Handbook of Statistical Tables. Addison-Wesley Publishing Co., Massachusetts. 1962.

Reyes-Castaileda, P. Diselio de Experimentos Aplicados. Editorial Trillas, Mexico. 1981. Rosner, B. Fundamentals of Biostatistics. Duxbury Press, Massachusetts. 1986.

Schefler, W.C. Biostatistics. Fondo Educativo Interamericano, Mexico. 1981.

Taguchi, G., System of Experimental Design, White Plains, NY, Kraus International Publications. (reprint of 1987 publication), 1994.

Vonesh, E., and V. Chinchilli, Linear and Nonlinear Models for the Analysis of Repeated Measurements, New York, Marcel Dekker, 1997.

Wilson, E., An Introduction to Scientific Research, revised edition, New York, Dover Publications, 1990.

Winer, B., D. Brown, and K. Michels, Statistical Principles in Experimental Design, 3rd edition, New York, McGraw-Hill, 1991.

Yates, F., The Design and Analysis of Factorial Experiments, Harpenden, England, Imperial Bureau of Soil Science, 1937.

Yates, F., Experimental Design: Selected Papers, London, Griffin Publishing, 1970.

Made in the USA
Monee, IL
04 November 2026

67378122R00118